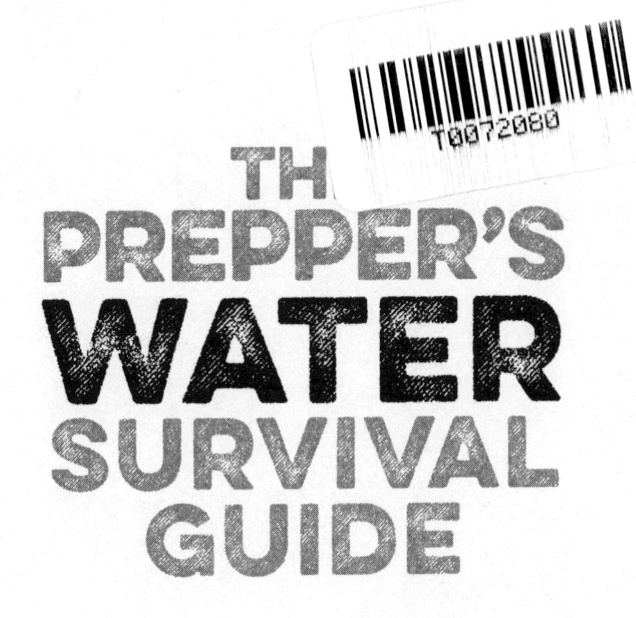
T0072080
TH
PREPPER'S
WATER
SURVIVAL
GUIDE

HARVEST, TREAT, AND STORE YOUR MOST VITAL RESOURCE

DAISY LUTHER

Published in the U.S. by
Ulysses Press
P.O. Box 3440
Berkeley, CA 94703
www.ulyssespress.com

ISBN: 978-1-61243-448-3
Library of Congress Control Number 2014952015

Printed in Canada by Marquis Book Printing

20 19 18 17 16 15 14 13 12 11 10 9 8

Acquisitions Editor: Kelly Reed
Managing Editor: Claire Chun
Editor: Renee Rutledge
Proofreader: Lauren Harrison
Indexer: Sayre Van Young
Cover design: Double R Design
Interior design: what!design @ whatweb.com
Layout: Jake Flaherty
Cover photos: plastic container © Designsstock/shutterstock.com;
water © Robert S./shutterstock.com

Thousands have lived without love, not one without water.

—W. H. Auden

CONTENTS

INTRODUCTION

If you've been prepping for a while, you've probably heard of the survivalist's "Rule of Three." You can survive:

Three minutes without air.

Three days without water.

Three weeks without food.

If a disaster has hit and you're still breathing, then your next concern has got to be water.

Have you ever watched any of those survival shows on the Discovery Channel, where people are dropped off in the middle of nowhere and left to survive with limited tools and supplies? In nearly every single episode, the biggest issue is finding and purifying water. Often, they wait so long that they become desperate and engage in risky behavior, like drinking water from a stagnant pool. In one particularly notable episode, the contestants had to be rescued because they became too weak from dehydration to seek water.

You don't have to be a contestant on a survival show or a survivor of a major disaster to require a water supply or a way to acquire it. There are a myriad of smaller issues that can spiral into a personal disaster if you don't have supplies on hand. What if:

- Your car broke down when you were driving through the desert and you had to wait or walk for help? Without water you could dehydrate very quickly in hot temperatures.
- You forgot or didn't have the money to pay the water bill and your utilities were cut off for a week?
- Your community was under a water restriction due to contamination of the water supply?
- The power went out and your home was on well water, thus halting your running water until the electricity was restored?
- You were out hiking and got lost, then were forced to spend a few nights in the woods with only the supplies in your daypack?

As you can see, those random occurrences that happen out of the blue can strike anyone at any time.

WHEN WATER IS LIMITED, CHAOS ERUPTS

It's easy to say, "Oh, I'll just go to the store and grab a few bottles," but when everyone else in your area has the same idea, it doesn't take long for the shelves to clear, potentially leaving you and your family without water.

Back in 2010, a water main broke in Boston, Massachusetts. The resulting leak flooded into the Charles River, and officials were forced to use the untreated backup reservoirs. A state of emergency was declared, a boil order was announced, and absolute chaos erupted as more than two million people suddenly found themselves without running water. A local news outlet reported:[1]

- The run on bottled water caused near panic at some stores throughout the Boston area Saturday night.
- At the BJ's in Revere, the crowd got so big and the rush for water so intense that police were called in. In order to maintain control of an unruly crowd, the store was shut down for the night.
- Shortly after residents in Boston received an emergency call warning them of the water crisis, supermarket aisles stocked with water were quickly wiped out.

1 http://www.wcvb.com/news/23407981/detail.html

- "They are fighting over it, literally fighting over water," said a customer at the Roche Bros. in West Roxbury. "I just had to fight my way through the aisles 'cause it's crazy in there."
- "Not since the Blizzard of '78 have I seen something like this," said the store manager. "New shipments that arrived were gone within seconds."
- In Coolidge Corner in Brookline, long lines formed at Trader Joe's, CVS, and Walgreens for any kind of bottled water, including sparkling and pricey designer bottles.

The governor of Massachusetts was able to lift the boil order a mere three days later, but during that short span, the National Guard was dispatched to deliver water, businesses were called upon to increase the water inventory brought to the local stores, and many restaurants were forced to close their doors due to the lack of safe drinking water.

YOU'RE GOING TO NEED MORE WATER THAN YOU THINK

Even if you are able to jostle your way to the front of the line and victoriously snag the last 24-pack of individual water bottles, if the situation lasts longer than expected, that paltry amount is not going to see you through it.

Why not? Because on average, the expected rate of consumption is 1 gallon per person per day. That doesn't include consumption for pets or what you'll use for sanitation. If the situation persists for more than a couple of days, you're going to need to bathe, clean countertops and floors, and wash dishes. Not only that, but you'll have to figure out a safe way to dispose of human waste.

The water that you store for your family should be enough to see all members of the household through a two-week period without running water. This is the bare minimum supply you should have on hand.

WHAT IF THE SITUATION PERSISTS FOR MORE THAN A FEW DAYS?

Sometimes, even an abundant stored water supply isn't enough. In more dire situations, water supplies can be interrupted indefinitely.

Do you remember the earthquake that devastated Haiti? That unexpected natural disaster took place in 2010, and some areas still do not have running water five years later. Five years. There's no way a person could store enough water to last for that long, so the people affected have had to completely change their way of life. They've had to learn how to acquire water for their needs, how to purify it so it doesn't make them sick, and how to conserve the limited amount they have available.

FINDING WATER ISN'T ENOUGH

Did you know that oftentimes, more people die in the aftermath of a disaster than in the disaster itself? And the number one cause of death? Contaminated water.

If you are thirsty—truly, desperately thirsty—it's human nature to drink whatever is available because your imminent demise from dehydration is more concerning to you than the pathogens in that dirty water you are gulping down.

But drinking contaminated water can lead to a host of dreaded diseases like dysentery, hepatitis A, viral gastroenteritis, cholera, shigellosis, typhoid, diphtheria, and polio. Just one person handling personal waste improperly can contaminate the water supply for hundreds, even thousands, of other people downstream from them.

FRESH WATER IS YOUR MOST VITAL PREP

Whether you are just getting started in the preparedness lifestyle or you've been at it for a long time, there's always something new to learn about water. There's just so much information about water that it deserves its own book. Aside from air, it is the most vital element of human survival. In this essential guide, you'll learn:

- How to store a substantial supply.
- How to acquire it in case your stores run out.
- How to make it safe to drink.
- How to combat what could be lurking in your water.
- How to conserve the water, because you have to make the water you acquire last until you can get more.
- Basic sanitation to keep you and your family safe and healthy.

What's more, a water supply and source aren't only important during disasters. It's vital to know about the things that could be lurking in your water even if it assumedly flows safely from your taps. Municipal water supplies and wells can contain things you'd rather not consume. Sometimes these contaminants are mild and only cause issues when consumed over a long period. Other times, the contaminants can make a susceptible person ill almost immediately.

There is nothing you can store that is more valuable than water or the means to purify water. There is no greater preparedness measure that you can take than that of securing a safe, abundant source of water. Without this one vital element that makes up 50 to 70 percent of your body, you're as good as dead. This could be the most important preparedness information you ever read.

CHAPTER ONE

EVEN NON-PREPPERS NEED WATER

We never know the worth of water till the well is dry.
—Thomas Fuller, *Gnomologia*, 1732

You don't have to be waiting for doomsday to see the need for a supply of water. An event that would be relatively minor if you had a water supply could turn into a true life-threatening emergency without one. Across the globe, unexpected predicaments happens, like:

- Terrible storms
- Chemical spills
- Car troubles
- Earthquakes

- The water to your home is shut off
- Droughts
- Wells running dry

With such situations, you rarely get any notice ahead of time. People aren't alerted to the fact that their car is going to break down later that day. No one is told, "Hey, there's going to be an earthquake tomorrow, so you'd better run to the store." You don't know beforehand that a tank is going to leak deadly chemicals into the municipal water supply. You just have to be ready in the event these things occur.

Sometimes people say things like, "You worry too much" or "You need to learn to relax."

What they're missing is this: Being prepared isn't pessimism. It's the ultimate act of optimism. It means that you are ensuring the survival and comfort of your family, no matter what may come your way.

IT WON'T HAPPEN TO ME

When interviewed after a crisis, most people say some variation of "I never thought this would happen to me" or "I never expected anything like this."

People never want to think about bad things striking close to home. But, as I mentioned before, we don't always get a warning, so maintaining a general state of preparedness can see you through all manner of things. In recent years, many disasters have occurred that made me want to up the ante on my own preparedness. In nearly all of them, the lack of clean potable water turned an event into a deadly crisis.

When you read the following scenarios, imagine yourself in the midst of them. Would you be ready?

OHIO, 2014

Time without Safe Running Water: 3 Days
Cause: Agricultural Run-off

Residents of the Toledo area were taken by surprise midsummer of 2014, when the local water supply became tainted and undrinkable. More than half a million people were affected. One moment, they had all of the fresh running water they could use. The next moment, it was announced this water could be deadly, and, suddenly, they had none.

Agricultural runoff had tainted Lake Erie, the city's water source, with a green slime called "algae bloom." Algae bloom is the sudden proliferation of microscopic algae in an aquatic system. The specific toxin in the Toledo case was microcystin. If consumed, symptoms like abnormal liver function, diarrhea, vomiting, nausea, numbness, and dizziness are likely to occur.

Microcystin affects animals even more than it does humans, so pets could not be given tap water, either. People complained of burns or rashes from showering or washing their hands in the water. Those with liver problems were particularly cautioned against having any contact with the water whatsoever.

People immediately flocked to stores to buy all of the water on the shelves. Stores in the vicinity reported that they were completely out of water within an hour of the announcement. Those who were able to traveled as far as 50 miles away to get drinking water. But what about those without transportation? What about those who did not have the money in their budget for a long drive and a big shopping trip? Some families were unable to purchase water and drank tainted tap water, becoming very ill with gastric distress.

The contamination was so bad that methods like boiling the water or using filtration units couldn't purify it. Boiling the water would actually intensify the toxins.

The National Guard delivered drinking water, but amounts were rationed and limited. Half a million homes were without drinking water for three days before the all-clear was given to begin consuming the tap water again.

WEST VIRGINIA, 2014

Time without Safe Running Water: 2 to 4 Weeks
Cause: Chemical Spill

In January of 2014, an industrial accident caused a deadly chemical to leach into the municipal water supply: 4-methylcyclohexane methanol, a toxic chemical used in the coal industry, leaked into the Elk River near Charleston, West Virginia.

The material safety data sheet (MSDS) is a document that contains detailed information on the potential hazards (health, fire, reactivity, and environmental) of and how to work safely with a variety of specific chemicals. It also advises first aid in the event that the chemical is ingested, inhaled, or comes into contact with someone's skin. The MSDS on 4-methylcyclohexane methanol[2] contains the following alarming warnings about the chemical that seeped into the water supply:

- Move out of dangerous area. Show this material safety data sheet to the doctor in attendance. Symptoms of poisoning may only appear several hours later. Do not leave the victim unattended.
- If inhaled: Move to fresh air. If unconscious, place in recovery position and seek medical advice. If symptoms persist, call a physician.
- In case of skin contact: If skin irritation persists, call a physician. If on skin, rinse well with water. If on clothes, remove clothes.
- In case of eye contact: Immediately flush eye(s) with plenty of water. Remove contact lenses. Protect unharmed eye. Keep eye wide open while rinsing. If eye irritation persists, consult a specialist.
- If swallowed: Keep respiratory tract clear. Do NOT induce vomiting. Do not give milk or alcoholic beverages. Never give anything by mouth to an unconscious person. Take victim immediately to hospital.

---

2 http://online.wsj.com/public/resources/documents/Eastman.pdf

When this chemical leaked into the Elk River, over half a million residents were affected. The toxin in the water was so deadly that a complete ban on tap water was issued. People were advised that the water was not safe to drink, bathe in, clean with, or cook with.

One reader of my website, *The Organic Prepper*, wrote and shared his experience:[3]

> *I have firsthand knowledge this water is not safe. For one, it smells very strongly of licorice. It has an off color, a blue hue. I will never drink it again. I felt it was probably OK to bathe in but that was a mistake, too. After flushing my system per the water company's specs, I took a shower. My eyes burned for four hours post-shower. Many other posts on Facebook and Twitter have complained of the same ailment along with rashes.*

Neither using water filters nor boiling the water removed enough of the chemical to make it safe. A statement from Berkey,[4] one of the most reputable filtration companies around, said:

> *The chemical that was leaked in West Virginia is an organic chemical. This is a relatively unknown chemical that is not on the EPA's organic chemical list to test for and for this reason we have not specifically tested this organic chemical compound, therefore we cannot say yes or no.*

---

3 http://www.theorganicprepper.ca/are-water-filters-effective-against-the-chemical-in-the-west-virginia-water-supply-01172014

4 http://www.theorganicprepper.ca/are-water-filters-effective-against-the-chemical-in-the-west-virginia-water-supply-01172014

Those area residents who did not have water stored (and there were many of them) were left to clear out store shelves, causing an immediate shortage. Those who were unable to purchase water locally had to drive to other towns to get drinking water. As the situation persisted, nearby towns were also cleaned out and many people had to make do with only what they had on hand.

The water ban lasted for 10 days, but many residents still complained of a red skin irritation not unlike a sunburn when they bathed in it, and most refused to drink the water, saying it smelled like licorice.

EAST COAST, 2012

Time without Safe Running Water: 2 Days to 2 Weeks

Cause: Hurricane Sandy

Hurricane Sandy, the so-called Frankenstorm that struck the East Coast in October of 2012, battered the highly populated areas around New York City and the New Jersey coast particularly badly.

After the terrible storm had come and gone, the danger was far from over. The aftermath was just as dangerous. The local sewage treatment plants discovered a very serious flaw in their setup: the electrical components were not above the flood level, and this released untreated sewage into the waterways.

And we're not talking about just a little bit of untreated sewage. We're talking about more than 10 billion gallons of human waste floating through area waterways.[5]

Floodwaters were deadly, and coming in contact with them put you at risk for infections and illness. The water contaminated all of the flooded homes, but the crisis didn't end there. In some areas, tap water was under a boil order for weeks after the storm.

As if contamination in the streets and the lack of potable drinking water flowing from the taps was not enough for the highly populated area to contend with, another dire issue caused a whole new disaster. In many parts of the city, there was no running water at all. The lack of running water meant that toilets could not be flushed. In a high-rise apartment building housing hundreds of people, that's very bad news.

Toilets backed up throughout buildings. Elderly people stranded on high floors had to contend with not only their own waste, but waste that erupted through their toilets from the apartments on floors below them. There were reports of people urinating and defecating in hallways in order to try to keep their homes livable. It took extensive sanitation measures during the cleanup process to make these areas safely habitable again.

---

5 http://www.nydailynews.com/new-york/report-sandy-filled-waterways-poop-article-1.1330630

CALIFORNIA, 2010

Time without Safe Running Water: 2 Days

Cause: Lost in the Desert

It was a hot day in July when three women decided to take a road trip through the hottest place on earth. By hot, I mean a sweltering 125°F (52°C). Death Valley in California is the site of the hottest air temperature ever measured on the planet (135°F/57°C), and it's also the driest place on the continent of North America. Probably not where you want to run out of water, right?

Donna Cooper, Gina Cooper, and Jenny Leung got a lot more adventure than they bargained for when the GPS unit that they were relying on directed them down the wrong route. The women were lost in the most inhospitable place on the continent, and they had embarked on what was supposed to be a day trip with only four individual water bottles.

They drove around all day trying to find their way back to civilization, to no avail. When they finally stopped for the night, the fuel tank was nearly empty and they had only one water bottle remaining.

On their second day lost in the desert, they tried in vain to signal passing airplanes. They finished off the last remaining water bottle. They were all suffering from signs of dehydration. At this point, they were more than 200 miles off course in an area so isolated that days could pass without signs of traffic.

They decided to drive as far as they could to look for help or shelter. By some miracle, they managed to drive the car, sputtering on the fumes of the tank, and stumbled upon an abandoned trailer. Mercifully, the trailer still had water going to the hose in the yard. The women were able to drink from the hose and cover themselves with water to lower their body temperatures.

According to the officials who found them, all three women would have certainly died if they had not located the trailer. That hose was the only thing that kept them alive for the next three days in the blazing heat until help finally arrived.

BOSNIA, 1992

Time without Safe Running Water: 1 Year

Cause: Military Blockade

Selco is a legendary figure in the preparedness world. On his website, SHTFSchool.com, he talks about living through the Balkan wars, when his city was blockaded by the military for an entire year. During that time, all that residents of the city had was what they possessed when the wars began. There was no running water or electrical power and no grocery stores or trucks delivering supplies. They were completely on their own, with only the resources they had on hand.

Selco writes graphically on the importance of water. He discusses that the general rule of surviving for a few days without water is misleading. You might stay alive, but it could lead up to your death if you're without water for sanitation and drinking. One specific example stands out: becoming crippled with a fungal infection on your feet means that you are unable to run. In a scenario like the one he describes, a lack of mobility could become a death sentence.

Selco reports seeing many people die due to sanitation issues and bad water. He and his family collected rain in barrels and disinfected it. The nearby river was too polluted to use for drinking water, but many desperate people drank it anyway. This resulted in waterborne illness, and sometimes death.

DON'T LET A BAD SITUATION GET WORSE

They say that hindsight is 20/20, but foresight is even better. In all of these examples, some preparation could have made a world of difference.

You don't have to be a doomsday prepper to see that the need for water applies to everyone, regardless of political beliefs, race, financial and social statuses, or geographic location. Across the planet, bad situations have become even worse because people did not have the foresight to store water, discover alternative sources for water, and have water purification options ready.

If those in the path of Hurricane Sandy had stored drinking water and sanitation supplies, they would have had a much easier time enduring the aftermath of the storm. If the women in California had put a case of water bottles or a few gallon-size jugs in the trunk of their vehicle, they would have contended with extreme heat but avoided the added physical suffering of dehydration. Families in Bosnia with secondary water sources and filtration devices would have had one less concern while they spent a year struggling for survival.

Often, government agencies like the Federal Emergency Management Agency (FEMA) or the National Guard provide relief for citizens affected by a disaster. But their resources are limited, and because of this, supplies must be rationed out. Everyone gets a little, but no one gets quite enough. People stand in line for hours, twice a day, to get an MRE, which is an army-issued "meal ready to eat" of dehydrated food and an individual bottle of water. Would you want to survive on only two bottles of water per day? Would you want to listen to your children say, "I'm thirsty, Mommy," and have nothing to give them?

Water preparedness is actually pretty simple. It's the least expensive item to add to your preps, but it gets overlooked by many. Why?

Well, it isn't glamorous, like a vast collection of guns, ammo, and body armor. It doesn't provide you with an interesting variety, like a long-term food supply could. But, be it ever so humble, it is the one thing you can't survive without for more than a few days, and that makes safe water the most important prep of all.

CHAPTER TWO

DEHYDRATION CAN BE DEADLY

The first concern presented by a lack of water is dehydration. The human body loses fluids through sweating and urination. Even the vapor released when you exhale causes you to lose fluid. These fluids must be topped off frequently or you become dehydrated.

Technically speaking, dehydration is the state that occurs when you use or lose more fluid than you take in, and your body doesn't have enough water and other fluids to carry out its normal functions.[6] Your electrolytes are out of balance, which can lead to increasingly serious problems.

Symptoms of electrolyte imbalances include dizziness, fatigue, nausea (with or without vomiting), constipation, dry

---

6 http://www.mayoclinic.org/diseases-conditions/dehydration/basics/definition/con-20030056

mouth, dry skin, muscle weakness, stiff or aching joints, confusion, delirium, rapid heart rate, twitching, blood pressure changes, seizures, and convulsions.

By the time you notice you're thirsty, you are already mildly dehydrated. According to the Mayo Clinic, the symptoms of mild to moderate dehydration are:[7]

- Constipation
- Decreased urine output
- Dizziness or lightheadedness
- Dry, sticky mouth
- Dry skin
- Few or no tears when crying
- Headache
- No wet diapers for three hours for infants
- Sleepiness or tiredness—children are likely to be less active than usual
- Thirst

If the situation continues and the person is not given fluids, severe dehydration can occur, and this is a medical emergency. The symptoms of severe dehydration are:

- Extreme fussiness or sleepiness in infants and children; irritability and confusion in adults
- Extreme thirst

7 http://www.mayoclinic.org/diseases-conditions/dehydration/basics/symptoms/con-20030056

- Fever
- Little or no urination—any urine that is produced will be darker than normal
- Low blood pressure
- No tears when crying
- Rapid breathing
- Rapid heartbeat
- Shriveled and dry skin that lacks elasticity and doesn't "bounce back" when pinched into a fold
- Sunken eyes
- In infants, sunken fontanels (the soft spots on the top of a baby's head)
- Very dry mouth, skin, and mucus membranes
- In the most serious cases, delirium or unconsciousness

There are three major reasons that dehydration is so common during a disaster situation:

1. People must perform strenuous tasks, causing them to perspire and lose fluids.
2. People may be ill and losing fluids due to vomiting and diarrhea.
3. There may be a lack of safe drinking water available to replenish them.

By the time dehydration becomes severe, your performance will have already become impaired, and during a crisis, you need to be functioning at full capacity to deal with the situation.

Once you reach a certain level of dehydration, you will cease being ravenously thirsty and not be thirsty at all. Because of this, the best way to gauge dehydration is not through thirst, but through the color of your urine. The urine of a well-hydrated person is almost clear, while a person suffering from dehydration has dark-colored urine. The darker a person's urine, the more severe the level of dehydration.

Dehydration affects some people more than others. Children, people with chronic illnesses, pregnant or nursing women, and the elderly are more vulnerable to dehydration than healthy younger adults. Other contributing factors that can exacerbate dehydration are:

- High altitudes
- Hot weather
- Humidity
- Strenuous physical labor

DEHYDRATION-RELATED AILMENTS

Dehydration can lead to very serious side effects, including death. Following are the most common dehydration-related ailments.

Heat cramps: Heat cramps are painful, brief muscle cramps. Muscles may spasm or jerk involuntarily. Heat cramps can occur during exercise or work in a hot environment or begin a few hours following such activities.

Heat exhaustion: Often accompanied by dehydration, heat exhaustion is a heat-related illness that can occur after you've been exposed to high temperatures. There are two types of heat exhaustion.

- Water depletion: Signs include excessive thirst, weakness, headache, and loss of consciousness.
- Salt depletion: Signs include nausea and vomiting, muscle cramps, and dizziness.

Heat stroke: Heat stroke is the most serious form of heat injury and is considered a medical emergency. Heat stroke results from prolonged exposure to high temperatures—usually in combination with dehydration—which leads to failure of the body's temperature control system. The medical definition of heat stroke is a core body temperature greater than 105°F (41°C), with complications involving the central nervous system that occur after exposure to high temperatures. Other common symptoms include nausea, seizures, confusion, disorientation, and sometimes loss of consciousness or coma.

Dehydration can lead to other potentially lethal complications. The Mayo Clinic offers the following examples:[8]

Seizures: Electrolytes—such as potassium and sodium—help carry electrical signals from cell to cell. If your electrolytes are out of balance, the normal electrical messages can become mixed up, which can lead to involuntary muscle contractions, and sometimes, loss of consciousness.

Low blood volume (hypovolemic shock): This is one of the most serious, and sometimes life-threatening, complications of dehydration. It occurs when low blood volume causes a drop in blood pressure and a drop in the amount of oxygen in your body.

Swelling of the brain (cerebral edema): Sometimes, when you're taking in fluids again after being dehydrated, the body tries to pull too much water back into your cells. This can cause some cells to swell and rupture. The consequences are especially grave when brain cells are affected.

Kidney failure: This potentially life-threatening problem occurs when your kidneys are no longer able to remove excess fluids and waste from your blood.

Coma and death: When not treated promptly and appropriately, severe dehydration can be fatal.

---

8 http://www.mayoclinic.org/diseases-conditions/dehydration/basics/complications/con-20030056

HOW TO TREAT DEHYDRATION

People who are suffering from dehydration must replace fluids and electrolytes. The most common way to do this is through oral rehydration therapy (ORT). In extreme cases, fluids must be given intravenously. In a disaster situation, hospitals may not be readily available, so every effort should be made to prevent the situation from reaching that level of severity.

Humans cannot survive without electrolytes, which are minerals in your blood and other bodily fluids that carry an electric charge. They are important because they are what your cells (especially those in your nerves, heart, and muscles) use to maintain voltages across cell membranes and to carry electrical impulses (nerve impulses and muscle contractions) across themselves and to other cells. Electrolytes, especially sodium, also help your body maintain its water balance.

Water itself does not contain electrolytes, but dehydration can cause serious electrolyte imbalances.

In most situations, avoid giving the dehydrated person salt tablets. Fresh, cool water is the best cure. In extreme temperatures or after very strenuous activities, electrolyte replacement drinks can be given. Sports drinks such as Gatorade can help replenish lost electrolytes. For children, rehydration beverages like Pedialyte can be helpful.

You can also make a homemade electrolyte drink. This way you know exactly what you are drinking, and it isn't a chemical cocktail like some commercial beverages might be. The powders can be mixed up ahead of time and tucked into a backpack, trunk of the car, or emergency bag, ready to add to water whenever they are needed.

These recipes were originally published in *The Prepper's Blueprint* and are used with permission from the author, Tess Pennington, who also owns the website ReadyNutrition.com.

HOMEMADE ELECTROLYTE DRINK

This option is made with sugar. When you work out, your body not only loses water and electrolytes, but burns energy as well. To make sure you can keep your activity level up, it is a good idea to add some kind of sugar to your drink.

2 quarts water (this can be added at the time of consumption to the dry ingredients below)

5 to 10 teaspoons sugar

1 teaspoon salt

1 teaspoon baking soda

½ teaspoon salt substitute (potassium salt)

1 individual packet sugar-free drink flavoring (like Crystal Light)

HOMEMADE ELECTROLYTE DRINK (SUGAR-FREE)

Although adding sugar to your drink will help you keep your energy levels up, it's not a good option for everyone. People on a low-carb diet or with diabetes can choose an electrolyte drink recipe that doesn't contain sugar.

VERSION 1

1 quart water

250 milliliters orange juice (citrus juice is a natural source of potassium ions)

3 tablespoons lemon juice

¾ teaspoon salt

VERSION 2

2 quarts water (this can be added at the time of consumption to the dry ingredients below)

1 teaspoon salt

1 teaspoon baking soda

½ teaspoon salt substitute (potassium salt)

1 individual packet sugar-free drink flavoring (like Crystal Light)

artificial sweetener, to taste

CHAPTER THREE

TOXINS IN MUNICIPAL WATER SUPPLIES

Gone are the days when we could simply drink water from natural sources without having to worry too much about what contaminants or pathogens lurked within. People have unwittingly consumed water infected with bacteria, viruses, and parasites, all natural sources of contamination for centuries. But advances in modern technology and industry, while beneficial to us in many ways, brought an unfortunate side effect: more water pollution.

Depending on where you live, the water flowing through your taps and faucets likely comes from one of two sources: groundwater or surface water.

There are problems inherent with each.

About 71 percent of the earth's surface is covered by water, and 97 percent of that water is saltwater. Only 3 percent is fresh water, and approximately two-thirds of that is frozen in glaciers and polar ice caps.

That means only 1 percent of all the water on earth is readily usable. Saltwater's sodium content is dangerous to humans; the salts in it are dehydrating. Your body would excrete more water than you consumed just to eliminate the salt content from your body via urination. Too much sodium taxes your kidneys and, because of this, drinking saltwater can ultimately kill you.

Saltwater can be desalinated, but the process is expensive and energy-intensive.

Fresh water is naturally occurring water on the earth's surface in ice sheets, ice caps, glaciers, icebergs, bogs, ponds, lakes, rivers, and streams, and underground as groundwater in aquifers (underground layer of water-bearing permeable rock or gravel, sand, or silt) and underground streams.

Drinking water usually comes from two main sources: groundwater and surface water (from lakes, rivers, creeks, and ponds).

Humans largely depend on groundwater. It supplies drinking water for 51 percent of the US total population and 99 percent of the rural population. It also helps us grow food: 64 percent of groundwater is used for irrigation to grow crops.[9]

---

9 http://www.groundwater.org/get-informed/basics/groundwater.html

The world's supply of groundwater is decreasing because the aquifers that underlie them are being depleted. Excessive pumping is a major cause of this depletion because it lowers the groundwater tables, placing them out of the reach of wells. As the water table lowers, it also becomes more expensive (sometimes even cost-prohibitive) to pump the water to the surface. Groundwater is replenished by rain and melted snow and is a source of recharge for lakes, rivers, and wetlands. It is brought to the surface by springs and can be discharged into lakes and streams. Because it is a source of recharge for lakes and streams, those bodies of water can also have their supply diminished when groundwater tables are lowered. In coastal areas, excessive pumping of groundwater can cause saltwater to move inland and upward, resulting in saltwater contamination of the water supply.

CONTAMINATION OF GROUNDWATER

Surface water is exposed to the environment and carries a higher risk of contamination than groundwater, which is naturally filtered as it passes through rock and sediment. But that doesn't mean groundwater is risk-free. It tends to be contaminated by naturally occurring arsenic. It also tends to be contaminated by man.

Gasoline, motor oil, road salts, mining site toxins, pesticides, herbicides, fertilizers, industrial and household chemicals, landfills, and untreated waste from septic tanks can all pollute groundwater.

Contaminants in drinking water supplies come from the following possible sources, according to the Centers for Disease Control and Prevention:[10]

- Chemicals and minerals that occur naturally, such as arsenic
- Viruses, bacteria, and parasites
- Local land-use practices, such as pesticide use
- Industry
- Sewer overflow and failing septic systems

With population increases come increases in bacteria, viruses, and related diseases. But human-caused water contamination started to become a problem during the Industrial Revolution that occurred from the 18th to 19th centuries, when factories began releasing pollutants directly into rivers and streams.

Despite the development of better waste management systems in the 20th century, the condition of our water did not improve much. An incident that occurred in Ohio in 1936 is a disheartening example of how industrial pollution has continued to destroy natural water resources in the US. Chemical waste was released into the Cuyahoga River, and a simple spark

10 http://www.cdc.gov/healthywater/drinking/public/water_diseases.html

from a blowtorch caused the waterway to burst into flames. The river caught fire several more times over the next three decades.

In 2007, CNN reported that "up to 500 million tons of heavy metals, solvents, and toxic sludge find their way into the global water supply every year."[11] In the developing world, as much as 70 percent of untreated industrial waste is dumped into rivers and lakes.

Water sources are also contaminated by rain runoff from such things as oil-slick roads, construction, mining, and dump sites. Livestock waste from farm operations, leaky septic tanks, pesticides, and fertilizers also taint our water supply.

Even natural spring water—once considered safe for all purposes—should now be tested before consumption.

In fact, the Environmental Protection Agency tells us to expect all drinking water—including bottled water—to contain at least small amounts of some contaminants.

What kinds of contaminants do they mean, exactly?

Here's a summary of the possible contaminants the agency lists on its website:[12]

- Microbes (bacteria, viruses, and parasites)
- Radionuclides (forms of radiation including alpha, beta, and photon emitters found in radioactive minerals, and radium)

---

11 http://edition.cnn.com/2007/world/asiapcf/12/17/eco.about.water

12 http://water.epa.gov/drink/contaminants/basicinformation/historical

- Inorganic contaminants/pesticides/herbicides (including but not limited to asbestos, cyanide, mercury, lead, fluoride, and arsenic)
- Volatile organic contaminants
- Disinfectants (chlorine, chloramine, chlorine dioxide)
- Disinfection byproducts (trihalomethanes, haloacetic acids, bromate, and chlorite)
- Radon
- MTBE (fuel oxygenates)

Some of these contaminants merely affect the taste of water, and others are tragically harmful to human health.

Some are actually beneficial to health, but that's usually in small doses.

Trace amounts of calcium, iron, and copper provide health benefits, but become dangerous in larger quantities. As Paracelsus, the founder of toxicology, said, "All substances are poisons; there is none which is not a poison. The right dose differentiates a poison."

The EPA has drinking water regulations for more than 90 contaminants. Currently, the agency is monitoring 30 contaminants that presently are not regulated, 28 of which are chemicals and two of which are viruses.

In late 2013, researchers from the US Geological Survey and the Environmental Protection Agency analyzed single samples of untreated and treated water from 25 US utilities that voluntarily participated in the project. They found traces of 18 unregulated chemicals, including 11 perfluorinated compounds, an herbicide, two solvents, caffeine, an antibacterial compound, a metal, and an antidepressant. The researchers said the concentrations were generally low, but for many of the contaminants, little is known about potential health risks. But one of the perfluorinated compounds, known as PFOA, has been linked to a variety of health problems, including cancer among people in communities where water is contaminated by a chemical plant in West Virginia.

PFOA has been detected in the blood of nearly all people in the US. Even more disturbing? A panel of scientists has concluded that there is a "probable link" between PFOA in drinking water and high cholesterol, ulcerative colitis, thyroid disease, testicular cancer, kidney cancer, and pregnancy-induced hypertension. The findings were based on people in Mid-Ohio Valley communities whose water was polluted with PFOA from a DuPont plant.[13]

IT DEPENDS ON WHERE YOU LIVE

The source of your drinking water plays a major role in determining its safety, and even that is often hard to estimate. Toxins present in municipal water supplies vary from city to city.

---

13 http://www.scientificamerican.com/article/unregulated-chemicals-found-in-drinking-water

In the Midwest, for example, there are high levels of pesticides (in particular, weed killer) due to agricultural practices that contaminate the groundwater (this also affects well water in the area). In 22 states with military contractors, perchlorate, the explosive component of rocket fuel, has been found in the tap water. In 2008, the Associated Press released a report informing the public that water treatment centers were unable to remove all traces of pharmaceutical drugs from the water supply. (The drugs were introduced into the water by human and animal urine.)

To determine the extent of drinking water contamination, an Associated Press investigative team surveyed the water providers of the 50 largest cities and 52 smaller communities in the United States. The team analyzed federal databases and scientific reports and interviewed government and corporate officials.

The investigation found widespread evidence of drinking water contaminated with both over-the-counter and prescription drugs, including painkillers, hormones, antibiotics, anti-convulsants, antidepressants, and drugs for cancer or heart disease.

Of the 28 major cities that tested their water supplies for pharmaceuticals, only two said those tests showed no pharmaceutical contamination. In Philadelphia, 56 different drugs and drug byproducts were found in treated drinking water, and 63 were found in the city's watershed.

THE SAFE DRINKING WATER ACT

If your primary water source is a public water system, it is regulated under the Safe Drinking Water Act. The EPA defines a public water system as one that serves 25 or more people per day. The utilities or companies that provide water this way normally collect water from a lake, river, reservoir, or underground aquifer, then treat it through the use of filters and chemical disinfectants before supplying it to customers through a pipe system. These providers are required to monitor their water for safety, and contaminants present must be kept at levels determined by the EPA. In addition, they are required to notify customers if a water safety issue arises.

The EPA's Consumer Confidence Rule requires public water suppliers that serve communities to provide consumer confidence reports (CCR) to their customers. Also known as annual water quality reports or drinking water quality reports, they provide summarized information regarding water sources used, detected contaminants, compliance, and educational information.

NITRATES: A SPECIAL CONCERN

Nitrates are another possible groundwater contaminant and are one of the most commonly found pollutants in rural areas.

Nitrate is an inorganic compound that contains nitrogen and water. Nitrogen occurs naturally in soil, but it also comes from synthetic sources. In nature, it comes from decomposing organic materials like plants and human and animal waste. When that organic matter decomposes, it returns nitrate to the soil.

Nitrogen-based fertilizers are a major source of groundwater contamination. This is due to runoff and overfertilization of crops: the nitrate that is not used by plants finds its way into water via precipitation, irrigation, and sandy soils.

Nitrate is colorless, odorless, and tasteless, so water testing is the only way to determine if it is in your water supply. Infants, pregnant women, nursing mothers, and elderly people are most at risk from nitrate contamination (even after short-term exposure), so water testing for those populations is highly recommended.

This is especially important for infants under six months of age, as they are particularly susceptible to nitrate poisoning. Here's why: nitrate is converted to nitrite in the body. Nitrite reduces the amount of oxygen in the baby's blood, causing shortness of breath and blueness of the skin (often called blue baby syndrome). The medical term for this condition is methemoglobinemia. This illness can cause the baby's health to deteriorate rapidly over a period of days and can lead to brain damage and death. Fortunately, if the illness is diagnosed early enough, it is treatable.

Nitrate itself does not usually cause health problems. It is the conversion to nitrite in the body, as mentioned above, that can cause illness. In significantly high quantities, nitrate that has converted to nitrite can form nitrosamines, which are known carcinogens (cancer-causing substances).

Little is known about other possible long-term chronic health effects from drinking water that contains high levels of nitrate.

The presence of nitrate in water is important for another reason: it is a strong indicator that other water quality problems like bacteria or pesticides exist. And, in a vicious cycle, bacterial contamination of water can increase an individual's susceptibility to nitrate. If your water source tests positive for nitrate, be sure to have it tested for other contaminants as well.

The maximum contaminant level, or EPA's drinking water standard, for nitrate is 10 milligrams per liter (mg/L), which is the same thing as 10 parts per million (ppm).

If a nitrate test shows levels higher than 10 ppm, you should find a safe alternative drinking water supply and take action to remove the nitrate from your affected water source before using it for consumption.[14]

---

14 http://www.epa.gov/region10/pdf/sites/yakimagw/faq_nitrate_and_drinking_water.pdf

PHARMACEUTICALS AND PERSONAL CARE PRODUCTS

As you can imagine, this category of water contaminants is expansive—the list of possible offenders is virtually endless.

The EPA defines pharmaceuticals and personal care products (PPCPs) as, in general, "any product used by individuals for personal health or cosmetic reasons or used by agribusiness to enhance growth or health of livestock. PPCPs comprise a diverse collection of thousands of chemical substances, including prescription and over-the-counter therapeutic drugs, veterinary drugs, fragrances, lotions, and cosmetics."

Some of the more concerning PPCPs that are most commonly found in our water supply are illicit drugs, antibiotics, steroids, chemicals used in agribusiness, residues from hospitals, and residues from pharmaceutical manufacturing.

You might be wondering exactly how some of these products are introduced to our environment. As most of us go about our daily lives, we probably don't think much about the impacts our personal routines have on our water supply.

But the truth is, even everyday household activities like showering, applying moisturizers, and cleaning our laundry, dishes, and homes can contribute to the contamination of our drinking water. Despite the best efforts of water treatment facilities, some of these contaminants are still measurable in the water supply. Here's how they get there:

- Oral medications and their breakdown products are excreted by human waste, flushed down our toilets, and introduced to our water supply through sewage or septic systems.
- Topical medications (those that are applied to the skin) and body-care products like shampoo, soap, and moisturizers are washed down our shower drains.
- Laundry and dishwashing detergents are sent down drains when we wash our clothing and dishes.
- Unused or expired medications are flushed down the sink or toilet, introducing them into the water supply.
- Unused or expired medications are thrown in the trash. When they get to the landfill and it rains, these can leach into groundwater supplies.
- Domestic animals create waste in yards (veterinary drug contamination). When it rains, the liquid combines with the waste to create toxic runoff that can contaminate our water supplies.

When it comes to the damage to our water supply, however, there are much bigger culprits than regular people going about their day-to-day lives.

Agribusiness and commercial operations also introduce a substantial amount of PPCPs into the water supply through various means, like:

- Excretion of veterinary drugs, including hormones and antibiotics, by farm animals into fields, where those chemicals run off into lakes and streams.
- Release of treated and untreated hospital waste (primarily acutely toxic drugs and diagnostic agents, not long-term medications) to domestic sewage systems.
- Disposal by pharmacies and physicians.
- Discharge of regulated industrial manufacturing waste streams.
- Disposal or release by underground drug labs and illicit drug users.

As you can see, it is pretty much impossible to prevent PPCP contamination of water. And, to make matters worse, technology to completely remove all PPCPs from water does not exist.

WHY ARE PPCPS A CONCERN?

A 2002 study by the US Geological Survey brought attention to PPCPs in water. Samples from 139 susceptible streams in 30 states were tested, and detectable (yet minute) quantities of PPCPs were found in 80 percent of the streams. The most common pharmaceuticals detected were steroids and nonprescription drugs. Antibiotics, prescription medications, detergents, fire retardants, pesticides, and natural and synthetic hormones were also found.[15]

To do your part to help lessen the contamination of your region's water supply by pharmaceuticals and medications, follow these guidelines for proper disposal:

- Flush prescription drugs down the toilet only if the label specifically instructs doing so.
- Dispose of unused prescription drugs through pharmaceutical take-back programs if available (check with your local pharmacy).
- Take unused, unneeded, or expired prescription drugs out of their original containers. Mix them with an undesirable substance, such as kitty litter or coffee grounds. Place the mixture in a sturdy, opaque, unlabeled container. Check for approved state or local collection programs or with area hazardous waste facilities for disposal. If there isn't one near you, place the container of the mixture in the trash.

An alarming 2006 study found detectable concentrations of 28 pharmaceutical compounds in sewage treatment plant effluents, surface water, and sediment. The drugs came from classes including antibiotics, analgesics, anti-inflammatories, lipid regulators, beta-blockers, anticonvulsants, and steroids.

Most of the chemical concentrations were detected at low levels, but the levels at which toxicity occurs is still unknown.

---

15 http://www.groundwater.org/get-informed/groundwater/products.html

How (or if) those chemicals are stored in the body, and their long-term effects are still a mystery as well. That's because, to date, studies on how long-term exposure to pharmaceuticals and personal care products in water impacts human health have not been conducted.

Absence of such research does not mean we shouldn't be concerned.

Studies have shown that exposure to many of those compounds do cause adverse effects in aquatic life. In their 2007 fact sheet on endocrine disruptors, the National Institute of Environmental Health Sciences (NIEHS) noted that "although limited scientific information is available on the potential adverse human health effects, concern arises because endocrine-disrupting chemicals, while present in the environment at very low levels, have been shown to have adverse effects in wildlife species, as well as in laboratory animals at low levels."[16]

Endocrine disruptors are naturally occurring or man-made substances that may mimic or interfere with the function of hormones in the body. They may turn on, shut off, or modify signals that hormones carry and thus affect the normal functions of tissues and organs. Endocrine disruptors can cause overproduction or underproduction of hormones. They may also interfere or block the way natural hormones and their receptors are made or controlled.

---

16 http://www.neiwpcc.org/ppcp/ppcp-concern.asp

The EPA currently has two lists of possible endocrine disruptors found in water. The first list, titled "Final List of Chemicals for Initial Tier 1 Screening," contains 67 items. The second list, titled "Final Second List of Chemicals for Tier 1 Screening," was updated in May 2014 and contains 109 items.

Some of the sources used by these providers are riskier than others: water coming from lakes and rivers is more likely to harbor problematic contaminants than water coming from underground aquifers. The number of households served by a provider also matters, as most violations occur at systems serving less than 10,000 people. The reason? Smaller providers often lack the resources to keep up with government standards and regulations.

The CDC lists the following as the top causes of disease from public water systems:[17]

- Giardia
- Legionella
- Norovirus
- Shigella
- Campylobacter
- Copper
- Salmonella
- Cryptosporidium

---

17 http://www.cdc.gov/healthywater/drinking/public/water-diseases.html

- E. coli
- Excess fluoride

If your primary source of water is a private well, you are not protected by the Safe Drinking Water Act. This means the responsibility for water safety testing (and necessary water treatment) falls upon you. All private wells use groundwater, which presents some of the same contaminant risks as public water sources do.

According to the CDC, the top causes of disease from well water are:[18]

- Giardia
- Campylobacter
- E. coli
- Shigella
- Cryptosporidium
- Salmonella
- Arsenic
- Gasoline
- Nitrate
- Phenol
- Selenium

While the Safe Drinking Water Act and improvements in sanitation and water treatment methods have helped to make

18 http://www.cdc.gov/healthywater/drinking/private/wells/diseases.html

our water safer, problems still exist. Some of those problems are caused by the treatment methods themselves.

Take the history of cholera, perhaps the best-known example of a disease transmitted via a bacterium that can infect water supplies. The infection causes over 100,000 deaths a year worldwide. There have been seven documented cholera pandemics attributing to tens of millions of deaths. The last outbreak in the United States occurred from 1910 to 1911.

Cholera is no longer considered a major health threat in countries with advanced water treatment and sanitation practices. But one of the methods used to kill cholera in a water supply is chlorination, which is great at removing bacteria, but doesn't come without a price: it creates numerous toxic byproducts, like chloroform and trihalomethanes. According to Dr. Michael J. Plewa, a genetic toxicology expert at the University of Illinois, chlorinated water is carcinogenic. "Individuals who consume chlorinated drinking water have an elevated risk of cancer of the bladder, stomach, pancreas, kidney, and rectum, as well as Hodgkin's and non-Hodgkin's lymphoma."[19]

Some public water systems add ammonia to chlorinated water, which poses even more risks. This "chloraminated" water is known to kill fish and reptiles, and studies on its impacts on human health are just beginning.

And there's another problem with chloraminated water—it reacts with the lead in water pipes, releasing yet another toxin

---

19 http://www.theorganicprepper.ca/huh-it-must-be-something-in-the-water-01082013

into the public water system. In Washington DC, when chloramination of the water first began, lead levels were found to be 4,800 times the UN's acceptable level for the toxic heavy metal.[20]

As you can see, commonly used water treatment methods sometimes transform compounds into new ones science currently knows even less about.

With so much uncertainty over government-regulated water purification methods, perhaps it is best to take matters into your own hands. If you must be reliant upon public sources, learning how to purify that water yourself may be your best bet.

ADDITIONAL SOURCES

www.cdc.gov/healthyhomes/bytopic/water.html
www.history.com/topics/water-and-air-pollution
www.oceanservice.noaa.gov/education/kits/pollution/02history.html
water.epa.gov/lawsregs/rulesregs/sdwa/ucmr/data.cfm

20 http://www.theorganicprepper.ca/the-great-american-genocide-07032013

CHAPTER FOUR

WATER-RELATED ILLNESSES

Pure water is the world's first and foremost medicine.
—Slovakian Proverb

When you're thirsty—truly thirsty to the point that your body is beginning to suffer the effects of dehydration—you will drink whatever water is available, even when you know that it will likely make you ill. The need to quench your thirst and hydrate your body will override your brain's warnings that the water isn't safe.

The number one cause of death in the aftermath of a disaster is waterborne illness, and there's a common mistake in treating it that can be deadly. It's extremely important to note that in some cases of water-related illness, antidiarrheal medications such as loperamide (Imodium) or diphenoxylate with atropine (Lomotil) can actually worsen the illness. Never

give anti-diarrheal medication if blood is present in the stool or if waterborne illness is suspected. It can actually worsen the symptoms in specific types of waterborne illness and should only be given after a definite diagnosis has been made. This is because many antidiarrheal products slow the rate at which food and waste products move through your intestines, allowing more time for your body to absorb the poisons that the bacteria is producing. This can greatly increase your level of illness and the risk of severe complications. Antinausea medications should also be avoided for the same reason. To quote Shrek, "Better out than in."

Following is a brief overview of these illnesses, along with their symptoms and treatments.

AMOEBIC DYSENTERY

Amoebic dysentery is the result of drinking water that is contaminated with harmful bacteria. In its milder forms, it's also known as "traveler's diarrhea" and "Montezuma's revenge" (although no one suffering from that would consider it mild!).

Amoebic dysentery is caused by the parasite Entamoeba histolytica. It is common in countries that use human feces as fertilizer for food crops. This can contaminate both food and water. Amoebas from an infected person form cysts, and those cysts cause severe inflammation in the intestinal tract. The onset of symptoms usually occurs two to four weeks after exposure.

SYMPTOMS

- Abdominal bloating
- Abdominal tenderness
- Chills
- Dehydration
- Explosive, often bloody diarrhea
- Extremely foul-smelling stool and flatulence
- Fatigue
- Frequent flatulence
- Moderate fever
- Severe abdominal cramps

Dehydration can cause further issues if fresh water is not available to rehydrate the person suffering from dysentery. In the most serious cases, the amoeba may travel to the liver and cause these additional complications:

- Enlargement of the liver
- Jaundice
- Pain in the upper-right quadrant of the abdomen

TREATMENT

Amoebic dysentery is treated with antimicrobial drugs. According to the CDC:

For symptomatic intestinal infection and extraintestinal disease, treatment with metronidazole or tinidazole should

be followed by treatment with iodoquinol or paromomycin. Asymptomatic patients infected with E. histolytica should also be treated with iodoquinol or paromomycin, because they can infect others and because 4 to 10 percent develop the disease within a year if left untreated.

In some cases, the symptoms of amoebic dysentery will go away without medication. Unfortunately, the amoeba that caused it will live on in the person's body. This could cause the diarrhea to return later and could cause infection to other people. Even when treated, symptoms can last for four to six weeks.

Remember, never give anti-diarrheal medication if blood is present in the stool or if waterborne illness is suspected.

DYSENTERY/SHIGELLOSIS

Dysentery (also called shigellosis) is caused by the bacteria Shigella. It is common in situations where a large number of people share an area with poor sanitation. After the Haiti earthquake, dysentery raced brutally through makeshift camps for survivors who had lost their homes.

SYMPTOMS

- Abdominal cramps
- Bloody stools
- Dehydration

- Fever
- Mucus in stools (sometimes a yellow mucus passes through portions of the intestinal membrane)
- Rectal tenesmus (this is a condition in which the victim feels as though they need to pass stool when none is present, resulting in straining, rectal cramping, and severe pain)

Careful sanitation and hygiene practices are vital to stop dysentery from spreading. The onset of shigellosis is 12 to 36 hours after exposure. It usually lasts four to seven days.

TREATMENT

Sometimes shigellosis is treated with antibiotics, particularly fluoroquinolones, ceftriaxone, and azithromycin. Many strains have become antibiotic-resistant and must run their course.

Remember, never give anti-diarrheal medication if blood is present in the stool or if waterborne illness is suspected.

E. COLI INFECTION

E. coli bacteria are naturally occurring and live in the intestinal tract of humans and animals. Most of the time, the bacteria is harmless or causes minor diarrhea. In the aftermath of a disaster, though, some particularly nasty strains (called "pathogenic") often crop up. Those strains are divided into the following categories:

- Diffusely adherent E. coli
- Enteroaggregative E. coli
- Enteroinvasive E. coli
- Enteropathogenic E. coli
- Shiga toxin-producing E. coli (Enterotoxigenic E. coli, or ETEC)

E. coli is transmitted through contact with fecal matter. This is why sanitation in the aftermath of a disaster is so incredibly important. Improper disposal of waste can taint water, making thousands of people ill. During a flood, the water can contain a huge amount of pathogens, so every effort should be made to avoid contact with floodwater. A person failing to wash his or her hands after a bowel movement can spread the disease as well, particularly if they are preparing food.

E. coli can also be contracted by eating meat that has not been cooked thoroughly. The bacteria can get into the meat during improper processing. Meat should be cooked to an internal temperature of 160°F (71°C) to prevent illness.

SYMPTOMS

- Diarrhea, which may range from mild and watery to severe and bloody
- Nausea and vomiting
- Severe abdominal pain and cramping

The symptoms of E. coli usually last for about a week. But, in some cases, particularly with the strain E. coli O157:H7, severe complications can occur about two weeks after the onset of diarrhea. These complications either signify hemolytic uremic syndrome (HUS) or thrombotic thrombocytopenic purpura (TTP). The symptoms of these significant complications include:

- Anemia
- Blindness
- Clotting disorders
- Coma
- Inability to move one side of the body
- Intellectual disability
- Irritability
- Long-term nervous system complications
- Low platelets in the bloodstream
- Pale skin (caused by anemia)
- Passing only small amounts of urine
- Renal (kidney) failure
- Seizures
- Slowness of movement
- Small dark patches or dots on the skin
- Tiredness or lack of energy
- Weakness or fatigue (caused by anemia)

In the most severe cases, E. coli can result in death.

TREATMENT

Milder cases of E. coli can be treated at home with oral rehydration therapy. It is vital that the patient not be treated with antidiarrheals, salicylates (aspirin, ibuprofen, or Pepto-Bismol), or antibiotics. Salicylates can increase the risk of intestinal bleeding. If the secondary symptoms listed above occur, this connotes a medical emergency and assistance should be sought if at all possible. Treatments may include:

- Blood transfusion to treat anemia by providing additional oxygen-rich red blood cells.
- Careful regulation of fluids and essential minerals.
- Dialysis to filter waste products from the blood. Some people with kidney failure caused by E. coli infection require dialysis.

HEPATITIS A

Hepatitis A is common in disasters that result in flooding. Well water supplies in particular can become contaminated by floodwater. Hepatitis A is a viral disease that comes from ingesting fecal matter. It can also be passed from person to person, particularly if the infected person handles food. Injectable drug use and sexual contact are other ways that hepatitis A can be transferred.

Hepatitis A is one of the easier waterborne illnesses to avoid. A filter won't remove it, but boiling water for three minutes will. Adequate chlorination will kill the virus as well.

Hepatitis, in general, is inflammation of the liver. Children are the most vulnerable to hepatitis A, but often show no symptoms, which means that they can transfer the illness more easily since no one realizes that they are infected.

SYMPTOMS

- Abdominal pain
- Abrupt onset of fever
- Fatigue
- Jaundice
- Joint pain
- Light, clay-colored stool
- Loss of appetite
- Nausea
- Weakness

Illness can last anywhere from a week to several months. Many people get vaccinated before traveling to a place with endemic hepatitis A.

TREATMENT

If you have been exposed to hepatitis A and can get medications within two weeks, immune globulins can prevent you from becoming ill.

If you are already ill, get plenty of rest, eat well, and drink lots of fluids from untainted sources. Remember that your liver is not functioning at full capacity, so don't do anything that will cause further stress. The liver is a filter that removes toxins from your bloodstream. Avoid fatty foods, alcohol, supplements, and over-the-counter medications during recovery. Once you have had hepatitis A, you will be immune from it in the future.

VIRAL GASTROENTERITIS

Also called the "stomach flu," viral gastroenteritis is miserable. It is especially prevalent during situations in which hygiene and sanitation are issues. The virus, once ingested, causes inflammation in the stomach and intestines.

SYMPTOMS

- Abdominal pain
- Abdominal cramping
- Chills, clammy skin
- Diarrhea
- Fever

- Joint and muscle pain
- Loss of appetite
- Nausea
- Perspiration
- Vomiting

TREATMENT

People can become severely dehydrated from viral gastroenteritis, and most treatments are related to the prevention or treatment of dehydration. Clear fluids and electrolyte replacements are vital. When food can be tolerated, it should be bland: toast, crackers, broth, rice, and bananas are recommended. If dehydration becomes severe, intravenous fluids may become necessary. Severe dehydration can occur very quickly in small children and the elderly. Be cautious about using antidiarrheal medications.

CHOLERA

Cholera is a severe bacterial infection of the small intestine. People can die within hours of the onset of symptoms. Death is caused by the rapid loss of fluids. The illness has been called the "blue death" because of the bluish color that the skin takes on due to the rapidity of the dehydration.

Cholera is contracted through eating food or drinking water that has been contaminated with the bacterium Vibrio cholerae. This is generally found in the feces of an infected person.

Cholera epidemics often occur in the aftermath of a disaster. Insufficient sanitation and disruption of water services are both contributing factors in crowded situations.

SYMPTOMS

- Dehydration
- Labored breathing
- Leg cramps
- Low blood pressure
- Profuse watery diarrhea
- Shock
- Vomiting
- Coma

TREATMENT

Less than 1 percent of people who are promptly rehydrated die of cholera. Sufferers should be given large amounts of oral rehydration solutions containing sugars and salts. Severe cases could require intravenous fluids. Antibiotics can aid in a speedy recovery but are far less important than quick rehydration.

TYPHOID

Typhoid, also known as "typhoid fever," is a potentially deadly illness caused by the bacterium Salmonella typhi. Infected people carry the bacteria in their bloodstreams and intestinal tracts. Some people continue to shed the bacteria, even after they recover. These people are called carriers, and since they don't show any symptoms, they can unknowingly infect large numbers.

Typhoid is transmitted through sewage contaminated with the bacteria, or through the consumption of food or beverages handled by someone with typhoid.

To prevent contracting typhoid when you are in an area in which the bacteria is present, the CDC says, "Boil it, cook it, peel it, or forget it." Avoid raw fruits and vegetables, bring water to a rolling boil for at least three minutes, and eat foods while they are still hot after having been thoroughly cooked.

SYMPTOMS

- Abdominal pain
- Chest congestion
- Enlarged spleen and liver
- Fatigue and weakness
- Headache
- Intestinal bleeding

- Loss of appetite
- Rash with flat, rose-colored spots
- Very high fever (often over 103°F/39°C)

TREATMENT

Typhoid must be treated with antibiotics, specifically fluoroquinolones, ceftriaxone, azithromycin, ampicillin, and trimethoprim/sulfamethoxazole. Recent issues with antibiotic resistance make typhoid harder to treat. It's important to note that you can still carry the bacteria even after you have physically recovered, potentially transmitting it to other people. This is particularly likely if you have not received antibiotics. If possible, you should have stool samples taken after recovery to see if the bacteria is still present in your body. Take precautions to keep from transmitting the illness to others until these tests come back free of the bacteria. Relapses of the illness can occur and are actually more likely in those treated with antibiotics.

LEPTOSPIROSIS

Leptospirosis is common in the aftermath of a flood. First a little biology lesson about rodents: they're incontinent and leak urine everywhere they go. Most rodents are carriers of leptospirosis.

Lizzie Bennett, author of the website *Underground Medic*, writes:[21]

> *Leptospirosis is already in a great many fresh water supplies. Any situation that puts humans in contact with fresh water increases the risk of infection. It is endemic in tropical and subtropical regions, but not purely due to the climate. Many of these areas are in the developing world, and as such these areas have less treated water, less refuse collection services, and no pest control systems beyond those employed by individuals. This is exactly the situation we will find ourselves in if there is any event, natural or man-made, that results in a long-term breakdown of the power grid, which results in water treatment plants going offline. Those currently living an off-grid lifestyle, or preparing to should some kind of collapse occur, should be aware of the dangers lurking in even the most pristine-looking natural water supplies.*

You may know leptospirosis as Weil's disease. It's a zoonosis; that is, it can cross the species barrier from animals to humans. The animals themselves are asymptomatic—they show no signs of illness or disease. Leptospirosis is spread by infected urine, and, less commonly, by other bodily fluids. The spiral-shaped bacteria are safe when dry, but live for an extended length of time in damp and wet conditions. They can, if they have not been dry for too long, reactivate on contact with water. The prime sites for the bacteria to be found are in the water itself,

---

21 http://undergroundmedic.com/?p=95#sthash.ZgpA2t10.dpuf

alongside it on banks, under vegetation, and in mud, puddles, and in gardens where infected animals may have urinated.

SYMPTOMS

- Abdominal pain
- Chills
- Diarrhea
- Headache
- High fever
- Jaundice
- Muscle aches
- Rash
- Red eyes
- Vomiting

TREATMENT

Leptospirosis should be treated as early as possible with antibiotics, preferably penicillin if the patient is able to take that drug. Those with extremely severe symptoms may require intravenous antibiotics.

CHAPTER FIVE

A GLIMPSE AT EVERYDAY LIFE WITHOUT RUNNING WATER

I had been a prepper for several years before the importance of water storage truly resonated. As with most lessons, I had to learn it the hard way (fortunately, for us, it was not quite as hard a lesson as those in Chapter One).

My youngest daughter and I spent a year living in a cabin in the woods in North Central Ontario. We lived in secluded splendor on the banks of a huge lake and on the edge of an enormous national forest.

We moved there straight from the city, so our "initiations" were many. In this breathtaking setting, we learned to provide our own heat and live without electricity and running water.

You know how it goes with the most important things you learn: we were thrown right into it and the choices were to either sink (by packing up our belongings and scurrying back to the city with our tails between our legs) or swim (by learning what we needed to in order to thrive in this drastically different environment).

The little cabin we moved to had electricity and running water supplied by a well 30 feet from our back door. Don't let these amenities fool you, though. Just because these things were present doesn't mean they were always available for our use. Try getting a heavy concrete well lid off when it's buried under 3 feet of snow, then get back to me about the ease of availability of the water within.

The first challenge came in the late fall when our well dropped to dizzyingly low levels. The drought that had been going on for years was not limited to the West Coast—even lush, green Ontario suffered from low water tables.

Everything that came out of the taps was cloudy and murky, with bits of sediment from the bottom of the well floating through it. This was certainly not acceptable for drinking, washing dishes, or cooking. We had a bit of water, but not enough to do a load of laundry, and even if we had enough, our clothes were unlikely to get very clean from that muddy-looking water.

This broke us in gently, since we could still flush and take a quick, if not particularly thorough, shower. This predicament inspired me to invest in a water dispenser for the kitchen and enough 5-gallon jugs to keep us supplied with drinking water for a month. We used this for drinking, cooking, and making ice.

Kitchen sanitation was difficult and with snow impending that would leave us stranded at the cabin, I was hesitant to use our newly acquired drinking water for cleaning purposes. Here are some of the solutions we came up with.

We dirtied as few dishes as possible. I cooked in the oven and lined the cast iron with tin foil, which could easily be disposed of. Then the cookware required only a quick wipe. We also ate from paper plates and used the same drinking glasses throughout the day.

I set up a reusable water filtration system for dishwashing. I didn't want to use my good Berkey filters for all of this sediment and gunk, so I rigged up a device for my faucet. I used a mesh sieve lined with a piece of fabric. (Flour sack towels work well for this, as do coffee filters for a disposable option.) I tied this little contraption to my faucet with a piece of garden twine and ran water through it slowly. It caught the greater part of the sediment. I put a splash of bleach and dish soap into the water. For rinsing, I used lake water that had been filtered through the Berkey, boiled, then seasoned with a little splash of bleach.

We cleaned with kitchen wipes. For wiping down counters, stoves, and food prep areas, we used antibacterial kitchen wipes. It seemed really silly to "clean" with dirty water. If it wasn't too dirty from the dishes, we used the dishwater for cleaning, too.

The issue with our well lasted for about three weeks before the rain began to fall. Water table levels rose, and with them, our well water levels. Soon the water was sweet and clear again and we felt we were over the hurdle of water shortages. Little did we realize that all of this was merely a warm-up for the big event that would soon be coming our way. It would give a whole new meaning to power outages.

We hadn't been there long when we had our first power outage. Being straight off the bus from the city, I thought that with a well and septic system, life would be easy in a situation without electricity. I was prepared for life without lights, appliances, or the Internet. But what I hadn't thought of was life without running water. Not even dirty running water. No. Water. At. All.

In the city, despite our other inconveniences, when the lights went out, the water still flowed from the taps and the toilets still flushed. Not so when you have a well run by an electric pump. I discovered this when I got a panicked cry from the bathroom.

"Moooooom! The toilet won't flush!" Rosie called.

I went in to see what was going on. "Wash your hands and I'll get it working again."

Rosie turned the taps to no avail. Uh oh. Then it dawned on me. The pump.

I sent my daughter down to the lake with a bucket to get water for flushing. We left the lid off the tank throughout the lights-out episode for the sake of ease.

Life without clean running water posed some problems, but it was nothing compared to life with NO running water at all. We rallied quickly. We had our 5-gallon jugs of drinking and cooking water. The lake had not yet frozen over, so we were able to haul up buckets of water for our other needs. But I was determined not to be stuck in this position later, when a foot of ice would cover the water of the lake.

I took notes throughout the outage, which was mercifully short, and came up with solutions to ease future electrical outages and their subsequent effects on water. More details on the nuts and bolts of water storage methods follow in Chapter Six.

- **Store tap water for sanitation**. We added to our water storage supplies by purchasing 1-gallon bottles of drinking water. After we consumed the drinking water from these, we refilled them with tap water. Many of these were stored near the bathroom.
- **Fill the bathtub as soon as the weather gets bad.** Because we lost power frequently, we began to immediately fill the tub the second the sky darkened and the wind began to howl. A full bathtub can provide a lot of flushes and washing-up water.
- **Place an old-fashioned pitcher and bowl on a stand to use for hand-washing**. For many solutions, you need only look as far as an antique shop to see what our ancestors did. We set up cups near the pitcher and bowl to use for brushing our teeth.

- **Order extra filters and parts for your water-filtration system**. Living in such a secluded area, the snow removal was not always dependable. With the filtration system, we never had to risk running out of potable water.
- **Use basins for dishwashing**. Instead of using the sink to wash our dishes, we used basins. This way the water could be reused for flushing after the dishes were washed.
- **Stock up on baby wipes**. When the power is out, you still want to keep clean. Baby wipes are a good way to take a quick sponge bath without using freezing cold water or using up your precious supplies. They can also be used for hand-washing and minor cleanups.
- **Keep a kettle on the woodstove**. Actually, keep a couple of them. This added moisture to the dry air in the cabin and had the benefit of hot water on demand for tea, cocoa, or for adding to the basin to wash up with.

By the time the next power outage rolled around, we had become pros, and it was barely a blip on our radar.

The real moral of this chapter is not what we did or how we did it, though. It is that you can't know what difficulties you will face without a practice run.

We're pretty good at living without running water now, because we have had a lot of practice. It's less convenient but it hardly feels like "roughing it" because we learned to resolve our issues. We discovered what we needed to make our lives more comfortable and we were able to stock up on those things at our leisure. Trust me, you don't want to discover you need something only to find that everyone else in your town has just discovered the same thing.

CHAPTER SIX

CREATING A WATER PLAN

Here's another famous prepper adage: "One is none. Two is one."

This applies to all matters of preparedness and water is no exception. If one method can fail, it often will, and generally at the worst possible time.

Having a backup means that if your first means of survival fails, you still have something to rely on. And having two backups—well, that's even better!

A sufficient supply of stored water will allow you the time you need to access and purify water without taking unsafe shortcuts. Try to keep a two- to four-week supply on hand for your family.

Considering how important water is to your survival, a four-layered approach is wise:

1. Store drinking water.
2. Have a way to acquire and purify it at your home.
3. Have a way to acquire and purify it if you are away from home.
4. Conserve water.

Just as no two families are the same, no two water plans are, either. Your plan will be dependent upon a wide array of variables, including:

- How many family members you have
- Ages and preexisting conditions of family members
- The water setup in your home
- Storage space
- Climate
- Natural resources on or near your property
- Personal finances

Believe it or not, the very first step to developing your water plan isn't to go and fill up any empty jugs you happen to have lying around. Instead, it's to sit down and take a long hard look at your situation. Be honest with your assessment. You won't do yourself any favors by overestimating the money you have available or the storage space you can allot.

HOW MUCH WATER DO YOU THINK YOU NEED?

To get a clear picture of how much water you actually need, you should try living without running water for a while. Nope, I'm not crazy. It's time to shut off your water and torture your family. It's time for a "No Running Water Drill."

You don't have to move to the boondocks of northern Ontario or a cabin in the woods. Throw down the gauntlet and challenge your family to a prepper's weekend extraordinaire. Just switch off the water to your home for a couple of days and see how you fare. Your kids may complain, but a single weekend without running water is highly unlikely to kill them.

Don't forget to take notes and make a list of what you'll need to stock up on for future outages. Carefully track your usage of the following:

- Drinking water
- Water you add to beverages (like brewed coffee or lemonade)
- Other beverages. In the event of a long-term water outage, all beverages will count toward your hydration. If you drink a soda pop, count it toward your water usage so that you don't shortchange yourself.
- Water used for cleaning (washing dishes, wiping counters, etc.)

- Water used for personal hygiene (brushing teeth, washing hands, taking sponge baths, etc.)
- Water used for pets
- Water used for cooking (pasta, rice, veggies, condensed soups, etc.)

You really need to do this for more than 24 hours to get a clear picture of your needs. One day without a shower isn't a big deal at all, but many people become uncomfortable after two or three days. Figure out how you'll resolve that and any other issues that arise by actually living it.

HOW MUCH WATER DO YOU ACTUALLY NEED?

There are a few general rules for water storage quantities. Standard prepper wisdom for water storage is:

- For drinking water, store 1 gallon, per person, per day. This will include enough water for cooking as well. Extreme hot weather, strenuous activity, illness, or conditions like pregnancy or diabetes are all variables that can increase your need for water.
- For pets, it is dependent upon their size. Generally speaking, if they are above 50 pounds, they'll need about as much as a human. If they are under 50 pounds, it may dwindle to half a gallon or less. Keep

in mind when rationing out the water that a pet's digestive system can handle water that would make a human very ill, so save the purified drinking water for the two-legged family members.

- For sanitation, a gallon per day, per person is ideal. The water you store for cleaning and personal hygiene doesn't need to be subject to the same stringent standards as drinking water.

Refer back to the "No Running Water Drill" that you performed with your family and compare the results to the numbers above. Are they pretty close, or is there a big gap between theory and reality? Generally speaking, it is wiser to go with the higher quantity, just to make sure you have your bases covered.

Now that you have your magic number, it's time to begin.

Once you've got a clear picture of how much water you need, you can start filling all of those containers! Read on for more information about safely storing water.

CHAPTER SEVEN

STORING WATER

As with any major undertaking, always start with the simplest step. In the event of a disaster, having stored drinking water means you can focus on other tasks initially, instead of worrying about finding potable water. Many people become extremely ill because their thirst makes them desperate enough to drink water that they know is questionable.

For our family, including pets and humans, the magic number is 11 gallons per day. This is a combination of drinking water and tap water. For a 30-day supply, we're talking about a whopping 330 gallons. Think about that amount for a moment. It definitely takes up some space in our basement, that's for sure.

What containers should you use for storing water?

There are all sorts of containers and systems out there for storing water. Your first inclination may be to use containers that you already have, and certain containers can provide an inexpensive start to your water storage. There are a few containers you should never use, no matter how carefully you clean them.

UNSAFE STORAGE

Milk or juice jugs. FEMA warns against reusing milk jugs for water storage. "If you decide to reuse plastic storage containers, choose 2-liter plastic soft drink bottles—not plastic jugs or cardboard containers that have had milk or juice in them. The reason is that milk protein or fruit sugars cannot be adequately removed from these containers and provide an environment for bacterial growth when water is stored in them."[22]

Containers that stored potentially toxic non-food items. Residue of non-food items can remain in the containers, no matter how carefully you clean them. This can taint your stored water with chemicals that may make you ill.

Containers that are not food grade. Food-grade plastic containers are marked "PET" or "PETE." Other types of plastic containers can leach toxic chemicals into your stored water, particularly if they are subjected to extremely hot or cold temperatures.

22 http://www.fema.gov/pdf/library/f&web.pdf

SAFE STORAGE

Two-liter soda bottles and gallon water bottles are excellent, low-cost options to get you started. If your family, like ours, doesn't drink soda, sometimes you can get empty 2-liter bottles from a friend or neighbor. We had neighbors who went through a 2-liter bottle every day, which helped us build our supply very quickly. If operational security, or OPSEC, is a concern, be prepared ahead of time with an answer for why you want all of those empty soda bottles.

If you want your water supply to last longer than a month, you may want to invest in a storage system for the sake of efficiency. It would take a lot of 2-liter soda pop bottles to create a six-month supply!

One of my favorite ways to store drinking water is in 5-gallon jugs. There are lots of options for larger jugs. Some are very reasonably priced, while others might be out of the thrifty prepper's budget. You can purchase the jugs on Amazon, at your local WalMart, or from a water retailer. They are made from various materials. The most commonly available are glass, polycarbonate, and BPA-free.

- The best-quality jugs are made of glass. Glass jugs don't contain potentially toxic plastic and chemicals. The downside is that these are prohibitively heavy as well as very breakable. If you live in an earthquake-prone area, glass jugs are probably not the most practical choice. A 5-gallon glass water bottle weighs approximately 20 pounds, so when it's full, it will weigh over 60 pounds. If the weight and fragility are not issues, glass water bottles are readily available from "brew-your-own" beer and wine shops.

- Polycarbonate bottles are the sturdiest of the plastics; however, they contain Bisphenol A (BPA), which has been proven to be an endocrine disruptor and is also suspected to be a carcinogen. Particularly if the water is exposed to extremely high or freezing temperatures, the chemical can leach into your stored water.
- Water bottles marked "BPA-free" do not have the harmful chemical Bisphenol A. The downside to these bottles is that they are very fragile and many people complain that they split easily. Because they are less sturdy, these bottles absolutely cannot be stacked, which means that they'll take up a lot more floor or shelf space than their sturdier counterparts.

If you are filling containers with tap water, FEMA offers these guidelines for storing it safely:[23]

- Thoroughly clean the bottles with dishwashing soap and water, and rinse completely so there is no residual soap.
- Additionally, for plastic soft drink bottles, sanitize the bottles by adding a solution of 1 teaspoon of non-scented liquid household chlorine bleach to a quart

23 http://www.ready.gov/water

(¼ gallon) of water. Swish the sanitizing solution in the bottle so that it touches all surfaces.

- After sanitizing the bottle, thoroughly rinse out the sanitizing solution with clean water.
- Fill the bottle to the top with regular tap water. (If your water utility company treats your tap water with chlorine, you do not need to add anything else to the water to keep it clean.) If the water you are using comes from a well or water source that is not treated with chlorine, add two drops of non-scented liquid household chlorine bleach to each gallon of water.
- Tightly close the container using the original cap. Be careful not to contaminate the cap by touching the inside of it with your fingers. Write the date on the outside of the container so that you know when you filled it.
- Store in a cool, dark place.
- Replace the water every six months if you are not using commercially bottled water.

I just want to reiterate this: it's of the utmost importance that you get NON-SCENTED bleach for your drinking water.

I disagree slightly with FEMA on the "shelf life" of water. It doesn't spoil if the container was thoroughly cleaned. If your water has been stored for longer than six months, you may wish to purify it before drinking it. (See Chapter Ten for purification instructions.) If your stored water tastes "flat," pour it back and forth from one container to the other to help reoxygenate it and improve the flavor.

While it's actually okay to carefully reuse containers for water storage, commercially bottled water (sealed) is a better choice if you can afford it. It lasts indefinitely without the need for further purification.

DISPENSERS

An important thing to keep in mind when making your selection is how heavy water is. One US gallon of water weighs approximately 8.5 pounds, so not including the container, a 5-gallon jug will weigh about 42 pounds. If you aren't physically able to maneuver something this heavy, you should look at 1-gallon containers or a large container that can be filled in place then dispensed into smaller containers. The most cost-effective method is a pushdown pump for the tops of bottles. These are available for only a few dollars.

For our basic supply (the one we rotate into use on a regular basis), we use 5-gallon jugs in conjunction with a water dispenser. If you use these bottles on a daily basis, you too may want to invest in a dispenser. A nice bonus to the plug-in dispenser is on-demand hot and cold water.

Avoid the bottom-loading dispensers. Although they're designed to make your life easier (you don't have to muscle a 40-pound water jug upside-down on top of the dispenser), they're dependent upon an electric-powered pump to pull up the water from the jug. However, the main purpose of your water storage is for emergency preparedness. Although they aren't as convenient, top-loading dispensers will work whether or not you have electrical power.

Another dispenser option is the countertop style. These dispensers are gravity fed and require no electricity. Options include stainless steel or the more decorative ceramic and porcelain.

THE WATERBOB

Sometimes we have a little bit of warning before a water emergency. For times like that, we have a gadget called a WaterBOB. It's a 100-gallon bladder that fits into your bathtub. If for any reason you suspect that you might lose running water, simply put the clean bladder into your tub, fill it from the faucet, and place the cap on it. It comes with a handy little pump so you can get water out when you need it. This is better than simply filling your bathtub, particularly if time is of the essence. You won't have to contend with soap scum or anything that might be lingering in your tub, and it has a handy cap to keep it clean.

MODULAR STORAGE SYSTEMS

For your secondary supply, many companies offer water storage systems for large amounts of water.

Some of these are modular, like the WaterBrick. Square, 3.5-gallon units can be stacked in different configurations for

WHY YOUR SWIMMING POOL IS NOT A VIABLE DRINKING WATER SUPPLY

Many people plan to rely on their swimming pools in the event of a water emergency. It's true that there you have about 20,000 gallons of pristine, treated water…initially. But, if the water emergency coincides with a power outage, the water won't stay clean for long. Without the pump running constantly, algae and bacteria will grow, tainting the water and making it unsafe for consumption.

Initially, the water should be safe to drink. The amount of chlorine in your swimming pool is generally not harmful. To reduce the overwhelming smell and flavor, simply leave the lid off of your drinking water container overnight and some of it will evaporate.

If you have a swimming pool in your backyard, in the event of a water emergency, immediately fill all of your clean containers with water from the pool and put a lid on them. After the first couple of days, you should not consume swimming pool water, but you could still use it for hygiene purposes and sanitation.

efficient storage. You can also purchase 55-gallon barrels for drinking water, but these are prohibitively heavy (over 450 pounds) and have to be filled where you intend to store them. A 320-gallon tank system is also available, if you have the space. Keep in mind that this is nearly a one-year supply of drinking water for a family member. The system stands about 88 inches tall (just over 7 feet) when fully assembled.

OUTDOOR WATER TANK

Another option is a large outdoor tank. These are very common in desert areas where water has to be trucked in. Look into a cistern or a 300-plus gallon tank. Keep in mind that weather fluctuations can damage the integrity of some plastic outdoor tanks. If you get freezing temperatures, be sure to leave some space in the tank for expansion. Better yet, search carefully for a weatherproof system.

CHAPTER EIGHT

ACQUIRING WATER

You've created a water plan and followed it. You've stored enough of the crystal-clear life-sustaining liquid to keep your family hydrated and meet your household's needs for several weeks.

But what if your supply runs dry? What if the unthinkable happens and you find you are down to the last few drops? You'll need to know where to find water in your area.

Well before the need arises, you should scope out your area to locate any possible sources of water you can tap into if necessary. This can include rivers, creeks, lakes, ponds, and springs. If you have yet to acquire your property, these are certainly things to look for when searching for a home.

One thing I hear a lot is people saying that they "have a plan" for water. I have to be harsh and give you a reality check.

If you haven't put that plan into action, laid the groundwork for safety by testing it, and perfected your acquisition methods, you have nothing but talk. It is imperative that you get your harvesting plans into place and begin using them.

Remember when we talked about doing a "No Running Water Drill" (page 75)? Its main purpose is to enable you to recognize any shortfalls in your plan before a situation requiring your water plan arises. It's exactly the same with your backup water source. If you aren't using it (at the very least for gardening, just to put things into practice) then you have no idea whether or not it will work.

The most ideal solutions for any prepper's homestead are either a deep well or a natural spring on your property. Those will keep you in stock of fresh, pure water indefinitely in most cases. Exceptions can arise if the groundwater is contaminated due to a natural disaster like an earthquake or flood, or man-made disasters like industrial accidents.

Take the time and make the investment to create safe water sources on your property. The first thing people think about is a well. It's easier than you might think to install this nearly boundless resource.

DRILLING A WELL

If your property doesn't already have a well, you can add one yourself. Before cities and towns built pipelines to supply water to homes, many people relied on wells as a water source.

It used to be that building a well was an arduous process, requiring hours of digging with a pick and shovel. Luckily (for your back and busy schedule), now there are easier methods you can use to build one on your property.

Of course, if you have the financial resources, you can pay someone to drill a well for you. The cost of having a well professionally installed can run anywhere from $1,500 to more than $10,000. This will depend on things like:

- The going rate in your area
- How deep they have to drill
- How difficult the drilling is (for example, will the driller be deterred by layers of stone?)

If that is prohibitive for you, or if you are the adventurous type and like to build things yourself, here's how.

Keep in mind that there might be legal codes specific to your region or underground infrastructure (pipes, electrical lines, sewage) that will keep you from implementing a system like this, so do your due diligence before you start digging. Check with your local health officials to find out about any regulations or permit requirements that may apply to your property.

CHOOSING A SPOT

Choose a location as far as possible (at least 50 feet) from septic tanks, sewer lines, animal shelters, and other potential contaminants. If your well is located too close to any of those, you'll risk contaminating your new water supply—or the groundwater source itself—which can mean infecting your neighbors' wells, too. Nothing says you're a terrible neighbor like contaminating someone's water supply, right?

Officials in your area likely have access to well logs and other geological information that can help you determine the best location for your well. They also can tell you what kind of elements you are likely to find beneath the surface once you start digging. Silt, sand, and decomposed granite are ideal for driven wells. Hard clay and rock might be very difficult or even impossible to dig through. Officials can also tell you how deep you'll need to dig to find water, and about the quality of the aquifer you'll be dealing with as well.

HOW DEEP?

If you are lucky enough to have a neighbor with a well, you can ask them for help with this. They may be able to give you an idea of how deep you are going to need to drill. If not, there's a quick way to test the depth you'll likely have to go to reach water: lower a weight on the end of a string down your neighbor's well and measure to the point where the string becomes wet.

Water is most commonly found in sandy layers. Going deep enough to reach one of those layers is key. Generally speaking, the deeper you go, the better quality the water you reach will be. You should start finding water at a depth of about 20 feet.

WHAT YOU'LL NEED

Driven wells are the fastest and easiest kind to create. This method uses a rocket-shaped device called a well point, which is a pipe with openings that allow water to enter (a well screen). You will need the following materials:

- Riser pipe. The riser pipe should be galvanized pipe in 5- or 6-foot lengths (with 6-inch nipple), which is easier for hand driving. A standard 21-foot length of pipe cut into four pieces normally is adequate for a driven well.
- Drive couplings. The drive couplings will be used to attach the pipe pieces together.
- Drive cap. You'll place the drive cap on top of the nipple that threads into the drive couple of the piece of pipe that you are driving. The cap's purpose is to protect the pipe threads from cracking as you drive the well point into the ground.
- Well point. A stainless steel well point is best, as minerals are less likely to build up on it.
- Post hole digger or soil auger

- Sledgehammer, large mallet or pipe driver
- Carpenter's level or plumb bob
- Pipe wrenches
- Pump (hand pitcher or electric)
- Weighted string (optional)
- Garden hose
- Wooden rod or closed-end pipe
- Check valve

You are going to be driving the well point down into the earth like a nail, so keep that in mind during the process.

DRIVING THE WELL

1. First, using the post hole digger or soil auger, dig a vertical hole at least 2 feet deep in the spot you have chosen for your well.
2. Rub a bar of soap over your well point. This coats the device, which will prevent debris from entering it, and will make it easier to drive into the ground.
3. Now, attach a length of the riser pipe to the well point with one of the couplings. Place the drive cap over the open end of the pipe. Make sure the connections are tight!
4. Next, place the well point with the attached pipe into the hole. Use your carpenter's level or plumb bob to make sure your well point/pipe is level.

5. Use a sledgehammer, mallet, or pipe driver to drive the well point/pipe into the hole. Be sure to strike on the drive cap!
6. When the drive cap is about 4 inches above the ground, unscrew it and add another section of pipe.
7. Rotate the piping clockwise periodically using a pipe wrench to ensure the couplings are tight.
8. When your pipe is about 10 feet deep, start testing to see if you've reached water yet by one (or both) of these methods:
   - Pour a gallon of water into the pipe. If the water disappears within 2 minutes, you have driven the pipe deep enough.
   - Listen for a hollow "bong" when you strike the pipe. If you hear that, you've reached the water table.
9. If you want to check to see how far into the water table you are, remove the cap from the pipe and drop in a weighted string. When it hits the bottom, pull the string back up and see how much of it is wet. To ensure good suction with your pump, it is important that the entire length of the well point be immersed, and preferably at least 2 feet beyond that to account for seasonal variations in the water table.
10. Continue the above procedure, adding sections of pipe as needed until you reach water.

11. Once your well is deep enough, you'll need to prepare to attach your pump.
12. Make sure enough of the pipe extends above the ground so you can attach your pump. A hand pitcher pump should be about 3 feet above the ground. An electric pump should be a foot or more above ground. This is to minimize the chance of well contamination in the event of flooding.
13. Before you attach your pump, clean your new well by surging it or flushing it with a garden hose. To surge it, take a wooden rod or closed-end pipe and rapidly work it up and down for about five minutes just below the water level in the well. This will draw fine, loose sand and silt into the well, leaving the coarser and more permeable material outside of your well point. Then you can remove the fine sand from the well with your pump. Or, flush the well out with a garden hose by jetting water into the pipe. Sand and silt particles will wash out around the hose.
14. You are ready to attach your pump. Do so according to the manufacturer's instructions.
15. Next, test to see how much water your well produces. It should be able to produce 5 gallons per minute of water flow. If it isn't producing enough, remove the pump, reattach the driver cap, and drive your well point/pipe deeper into the ground.

16. Your well water will be visibly dirty at first, but it shouldn't take long for it to produce clear, clean water. Be sure to connect a check valve below the pump to prevent the water from going back down into the well.

WELL MAINTENANCE AND PROTECTION

After installing your well, there are things you can do to protect it from damage and contamination.

- Avoid using pesticides, fertilizers, herbicides, or other chemicals around your well and the surrounding land.
- Slope the area around the well to help drain surface runoff away from it.
- If you remove the pump for any reason, place a well cap or sanitary seal over the opening to prevent debris or pollutants from getting in.
- Keep accurate records of any maintenance, such as disinfection or sediment removal, that may require the use of chemicals in your well.
- Do not dispose of waste or chemicals in dry or abandoned wells.

Remember, the responsibility for testing well water for contaminants lies with you. Be sure to do this on a regular basis. Have the well tested once a year for coliform bacteria—nitrates at the least. Of course, you should be filtering and purifying your well water before consuming it, anyway, but it is good to know what you are dealing with. (Go back to Chapter Three for more information on the potential contaminants in well water.)

WHAT IF A WELL IS NOT AN OPTION?

If your property does not have a well and you do not want to install one (or can't install one because of local legislation), you will need to find other natural sources of water in your area.

SPRINGS

Springs occur when water pressure causes a natural flow of groundwater onto the earth's surface. They also can discharge fresh groundwater either directly into the beds of rivers or streams, or into the ocean below sea level.

So, a spring is a site where the aquifer surface meets the ground surface. As rainwater enters or "recharges" the aquifer, pressure is placed on the water already present. This pressure moves water through the cracks and tunnels within the aquifer, and sometimes this water flows out naturally to the surface.

Springs typically occur along hillsides, low-lying areas, or at the base of slopes.

Usually, spring water is remarkably clear. In some locations, however, the water is tea-colored. Even though it looks "dirty," it may actually be fine to drink. The reason for this tinge of brown? Natural tannic acids from organic material in subsurface rocks discolor the water.

Springs can be a reliable and relatively inexpensive source of drinking water if they are developed and maintained properly.

If you are fortunate enough to have a spring near you, here's what you need to know about collecting and consuming its waters.

SPRING WATER QUALITY

Springs invoke imagery of cool, crystal-clear, refreshing water. In ancient times, they were viewed with much intrigue and fascination, and were often believed to have healing powers. But looks can be deceiving.

It is important to understand that even if spring water looks pristine, chances are it isn't.

In fact, spring water quality can vary greatly because of factors such as the quality of the water that recharges the aquifer and the types of rocks with which the groundwater is in contact. The rate of flow and the length of the flow path through the aquifer affect the amount of time the water is in contact with the rock, and thus, the amount of minerals that the water can dissolve. The quality of the water also can be affected by the mixing of freshwater with pockets of ancient seawater in the aquifer or with modern seawater along an ocean coast.

SO, YOU FOUND A LOCAL SPRING—NOW WHAT?

There are a few signs you can look for to help you determine if a spring will be a reliable source of water for your family.

One of the key signs of a good spring is that the water maintains a constant temperature throughout the day. That temperature should be just below the average air temperature. Color is also a possible indicator of quality: the water should be crystal-clear. It is also important to consider the rate of flow: Does the water flow fluctuate throughout the year, or is it consistent? If the flow fluctuates greatly, that can be an indication that the source is unreliable or that the water has the potential to be contaminated.

Note the flow rate of your spring during late summer and fall because that's when groundwater levels and spring flows are usually at their lowest. If you are going to rely on a spring for drinking water, it should supply at least 2 gallons per minute throughout the entire year unless you have another reliable source of water to fall back on.

To test the flow rate of a spring, firmly place a 5-gallon bucket into the slope of the spring and allow the water to flow into the bucket. Track how long it takes the water to fill the bucket.

If the spring you located seems like it will be a consistent source, great!

But don't get too excited yet. Remember, any natural water source should be considered risky until proven otherwise.

Water from springs, no matter how perfect it looks, is no exception.

POSSIBLE CONTAMINANTS IN SPRING WATER

Unfortunately, spring water is susceptible to contamination through various means, just like other natural water sources are.

Microbes, viruses, and bacteria are not visible to the naked eye, so it is important to test and purify spring water before consumption. I know I keep saying that, but that's because it can't be emphasized enough. Waterborne illnesses can do more than give you an upset stomach—they can be deadly!

Here are some of the possible contaminants that could be lurking in your spring.

COLIFORM BACTERIA

Coliform is a family of bacteria commonly found in soils, plants, and animals. The coliform family is made up of several groups, one being the fecal coliform group, which are universally present in the intestinal tracts of all warm-blooded animals, including humans. The presence of fecal coliform in water is evidence that human or animal waste has been or is present.

The presence of some fecal material in lakes, ponds, and rivers is to be expected as part of the environment. After all, we share this earth with many creatures.

While coliform bacteria itself isn't likely to cause illness, its presence in water sources that are used for consumption can be a warning sign. That's because where there's fecal matter, there are likely other dangers. Many of the pathogens that are found in water originate in human or animal feces, including parasites, protozoa, viruses, and disease-causing bacteria.

It is relatively easy and inexpensive to test water for coliform bacteria, and because their presence often means other pathogens are likely infecting a water supply, coliforms are often referred to as "indicator organisms."

There are three different groups of coliform bacteria: total coliform, fecal coliform, and E. coli. Each has a different level of risk.

Total coliform bacteria. Commonly found in the environment (soil, vegetation, water), this type of bacterium is generally harmless. If only total coliform bacteria are detected in drinking water, the source is likely environmental and fecal contamination is not likely. However, this doesn't mean all is safe and clear: if environmental contamination can enter the system, there may also be a way for pathogens to enter the system. Therefore, it is important to find the source and resolve the problem.

Fecal coliform bacteria. A subgroup of total coliform bacteria, fecal coliform bacteria appear in great quantities in the intestines and feces of humans and animals. The presence of fecal coliform in a drinking water sample often indicates recent fecal contamination, meaning that there is a greater risk that pathogens are present than if only total coliform bacteria is detected.

E. coli. E. coli is a subgroup of the fecal coliform group, and possibly the most infamous. Most E. coli bacteria are harmless, found in great quantities in the intestines of humans and warm-blooded animals. As most people know, some strains—E. coli O157:H7 in particular—can cause serious illness. The presence of E. coli in a drinking water sample almost always indicates recent fecal contamination, meaning there is a greater risk that other pathogens are present.

The presence of any of these types of coliform bacteria in your water supply means you should not consume the water until action is taken.

It means something is amiss, and you need to determine what that is.

Groundwater in a properly constructed well or spring should not contain any type of coliform bacteria. If coliform IS found in a well or spring, it generally means that surface water has somehow leaked into the water. There are many ways this could happen: rain runoff or snowmelt finding its way into the well or spring through cracks, gravelly soil, or sandy soil. It could also be due to poor construction or cracks in the well or spring casing.

Insects, snakes, mice, or other wildlife can also introduce harmful pathogens into springs.

NITRATES

Another common—and concerning—spring water contaminant is nitrate-nitrogen. This typically comes from sewage, animal waste, or nitrogen-based fertilizers leaching into water.

Common sources of spring water contamination are septic systems, farms, fertilizers, pesticides, chemical or petroleum leaks, dumps, and landfills.

If you are going to rely on a local spring for drinking water, be sure to test it for contaminants on a regular basis. Late spring and early summer are generally the best times to test your water, since that's when coliform bacteria are most likely to show up.

As with water from any other source, ideally you will use purification methods before consumption, no matter what test results show.

RIVERS, LAKES, PONDS, STREAMS, AND CREEKS

These bodies of water are everywhere and would seem to be the source of an endless supply of water, should you have access to them.

But they, too, carry risks, and plenty of them.

It is safe to assume that water from rivers, lakes, ponds, streams, and creeks is, well, NOT safe to drink, as the possibilities for contaminants are virtually endless.

Untreated surface water can harbor all kinds of things you don't want your family to consume, including bacteria, viruses, parasites, chemicals, pesticides, gasoline, oil, heavy metals…the list goes on and on.

Because surface water is "live" in that it is constantly moving and changing, it is unpredictable. And contaminants that weren't there yesterday could very likely be there today; it is impossible to tell when a new pollutant has come along and infected any of these bodies of water.

If you want to take advantage of any of these sources of water, you'll need to be sure you follow the proper filtration and purification methods. (More about those methods is coming up.)

RAINWATER

If your property doesn't have any bodies of water nearby, you can use a catchment system for rainwater. This is the ultimate DIY water source. If you're lucky enough to live in a place with substantial rainfall, even if you're in the middle of a city with no other natural sources, you've got it made.

Rainwater has long been valued for its purity and softness. It has a nearly neutral pH and is free from disinfection byproducts, salts, minerals, and other natural and man-made contaminants. Rainwater harvesting is an ancient technique, but its popularity as a renewable, sustainable, and high-quality source of water has been increasing in recent years.

Most rainwater is pure until it comes in contact with surfaces that are not. However, if you live in an area that produces heavy industrial pollution, your rainwater may contain some undesirable contaminants. You can find out if this is a risk by contacting your local municipal government before you set up a harvesting system.

If you get frequent precipitation throughout the year, numerous water butts or barrels at the corners of your structures may supply enough water for your needs, including supplementing your garden. (Be warned that the "eco-police" in some places believe that the government owns the water falling from the sky; rainwater catchment is actually illegal in some states.)

Rainwater harvesting may not seem like it would yield much, but you'd be surprised: According to some estimates, for every inch of rain that falls on a catchment area of 1,000 square feet, you can expect to collect approximately 600 gallons of rainwater. Ten inches of rain falling on a 1,000-square-foot catchment area will generate about 6,000 gallons of rainwater—6,000 gallons! If you think you are one of the lucky people who will be able to collect that much (or more), be sure you have large cisterns or other vessels ready.

Don't let the thought of having to do calculations deter you. Setting up a rainwater collection and storage system is fairly easy, and there are various ways to do it.

First, you'll need to find areas around your homestead where you can collect water. You can collect runoff from anywhere the rain doesn't soak into the ground. This can mean the roof of your home or any other structures on your property.

Keep in mind that any catchment area is going to catch a lot more than water. Leaves, twigs, animal droppings, and other natural debris will likely get mixed in with your water. As with any water you are going to use for consumption, you'll need to filter it before drinking it.

There are some types of roofs that are not recommended for water-harvesting catchments because the risk of contamination is too high. Old tar and gravel roofs and those made with asbestos shingles or treated cedar are some examples.

So, if you don't know what materials your roof is made of, it is important to find out prior to using it as a catchment system.

You'll also need to know what your gutters are composed of. If they contain lead soldering or are coated in lead-based paint, you won't want to use them to collect water that your family will be drinking. It's just too risky. Replace them with gutters made of non-toxic materials if you'd like to use them to harvest rainwater.

SETTING UP A RAINWATER HARVESTING SYSTEM

Are you ready to set up your harvesting system? Here's how to get started.

The size and complexity of your rainwater harvesting system is up to you, but no matter how big you decide to go, your system will consist of all or most of the following components:

1. A catchment surface is the collection surface from which rainfall runs off.
2. Gutters and downspouts channel water from the roof to the tank.
3. Leaf screens and first-flush diverters are components that remove debris and dust from the captured rainwater before it reaches the tank or barrel.
4. One or more storage tanks, barrels, or cisterns catch the water.

HOW MUCH CAN YOU COLLECT?

If you want to get an idea of how much rainwater you will be able to collect, here's how to do that calculation.

First, calculate the square feet of your house's catchment area. To do this, measure the area of the outside walls and then include the overhang of any eaves. For example, let's say you have an oblong house with outside dimensions of 36 feet by 46 feet. You've calculated the overhang of your eaves as 2 feet. So, add the 4 feet of the eaves to each wall length (2 eaves of 2 feet equals an additional 4 feet for each wall) to get the total length of the walls plus the eaves (40 by 50 feet).

Now multiply 40 x 50 (length x width) to get your total roof catchment area:

(36 + 4) x (46 + 4) = 2,000 square feet

Since 1 inch of rainfall provides approximately 600 gallons of water for a 1,000-square-foot catchment area and the house in our example has a 2,000-square-foot catchment area (twice the area), you will multiply 600 gallons by 2:

600 gallons x 2 = 1,200 gallons

This means that if you have an average rainfall of 20 inches per year, you have the potential to collect 24,000 gallons of water in one year.

That's amazing.

5. A spigot and overflow valve deliver the water.
6. Filters and other methods treat and purify the water, making it safe to drink.

You can build your own using the items listed above or purchase a rainwater harvesting system or kit. Some kits come with all of the components you'll need, from the downspouts all the way to the barrel (some even have spigots), and some contain all or most of the parts, but not the barrel.

Of course, the treatment and purification component won't be built into your rainwater harvesting system, but it is so important that I included on this list. You'll learn about water purification methods later in this book.

Whether you are going to build your system from component parts yourself or use a kit, you'll need to take a look at your gutter system first to see what you'll need.

GUTTERS AND DOWNSPOUTS

If you already have a downspout on your house, you may need to disassemble it to reroute it to your barrel. Most downspouts go all the way to the ground, so you'll have to remove sections or assemble a new elbow downspout or a downspout redirector or diverter, which are products that are specifically designed for rainwater harvesting. Remember to use components that do not contain lead, as it can leach into your water.

LEAF SCREENS AND FIRST-FLUSH DIVERTERS

Keeping as much debris as possible out of your stored water by installing leaf screens will save you from having to deal with constant clogging and backups in your gutter and downspout. You can place leaf screens along your gutter, inside of the downspout, or on the end of the downspout. Putting leaf screens in all of those areas provides several catch points, which is even better.

First-flush diverters aren't necessary, but they are another way to stop debris from entering your storage containers. These T-shaped devices have a ball-and-seat system that flushes off the first water of a storm before it enters your barrels. This is the water that could be the most contaminated by particulates, bird droppings, and other materials laying on the roof.

ONE OR MORE STORAGE TANKS, BARRELS, OR CISTERNS

If you are planning to use your collected rainwater for human or animal consumption, the barrels or cisterns you use must be food-grade. Also, be sure the holding vessels you select are designed to hold water. They must be opaque to help inhibit algae growth.

Vessels you can use to catch and store the water come in various shapes, sizes, and styles, including simple barrels, tanks, and decorative urns. As previously mentioned, some of these storage containers have built-in spigots or taps.

Most water-storage barrels and urns come with covers. It is absolutely necessary to be sure your storage container, whatever kind it is, has a sealed cover. This will prevent people (children are especially at risk) and animals from falling in and drowning.

Also, covers prevent you from accidentally creating a mosquito-breeding ground. The blood-sucking disease vectors love standing water and will view uncovered water vessels as an invitation to paradise. Any small openings (around the downspout, for example) should be covered with a fine screen to prevent the critters from being able to access your water.

If you don't already have a flat, hard surface to place your barrels on, you'll need to find or build a flat, sturdy platform or place them on cinder blocks. Dig a 5-inch-deep rectangle in the area you've selected to place your barrels. Fill it with a layer of pea gravel and place your platform or cement blocks on top of it. This will provide better drainage around your rain barrels and help keep water away from the foundation of your home.

SPIGOT AND OVERFLOW VALVE

If you purchased or already own a barrel or urn that has a built-in spigot or faucet, you are set. If your storage vessels don't have a spigot, you'll need to install one yourself. Do this before you install the downspout. Here's how:

1. Drill a spigot hole into the side of your barrel. It should be high enough up on the barrel to fit a bucket or water jug underneath. Make a ¾-inch hole to properly fit the spigot you bought. (This is the standard size for a spigot; if you're using a different sized–spigot, make sure you drill the right-sized hole so that it fits into the side of the barrel.)
2. Squeeze a circle of caulk around the hole. Put caulk on both the inside and the outside of the barrel.

3. Attach the spigot. Put the spigot and the coupling together. Use Teflon tape to wrap the threaded ends to create a tight seal and prevent leakage. Put a washer on the threaded end of the coupling and insert it through the hole in the barrel from the outside. Slip another washer over the pipe from the inside. Attach the bushing to hold the spigot in place.
4. Follow the directions for attaching the type of spigot you have. You may need to attach it in a different manner than what is specified here.

Next, make an overflow valve:

1. Drill a second hole a few inches from the top of the barrel. It should be ¾ of an inch, or the same size as the first hole you drilled.
2. Squeeze a circle of caulk around the hole, both inside and outside the barrel.
3. Place a washer on the hose adapter and put it through the hole from the outside. Put another washer on the inside threads, attach some Teflon tape, and attach a nut to tighten the assembly. You can attach a garden hose directly to the valve if you ever need to.

You can connect multiple barrels together to make an overflow system. To do so, drill a third hole in the first barrel. Drill it at the same level as the spigot several inches to the side. Then drill a ¾ inch hole in the second barrel at the same level as the hole you just drilled in the first one. Attach hose adapters to the holes in both barrels as described above.

If you're using a third overflow barrel and customizing your system with a spigot in each barrel, the second barrel will need a third hole so you can connect it to the third barrel. Make a second valve on the opposite side of the barrel at the same level. Make a valve in the third barrel as well.

MAINTENANCE TIPS

It is very important to make sure your barrels are situated on study, level surfaces. The more full they get, the heavier they get, and you don't want them tipping over onto anything, or worse yet, anyone (including yourself).

If you live in a region where temperatures regularly reach freezing (or below), it is probably best to disconnect your rainwater harvesting system during the winter. This is because continuous freezing and thawing of the water in your barrels or urns can damage them.

Also, watch for overflow. If you get an excessive amount of rain or stop monitoring the volume of water in your barrels, they may overflow and cause damage to your home's foundation.

SMALLER-SCALE RAINWATER COLLECTION DEVICES

If you do not have the space to build a rainwater system that uses barrels, you can collect rainwater in smaller amounts using devices like RainSaucers. They look a bit like an inside-out umbrella and are made out of food-safe fabric.

RainSaucers are portable and currently available in three sizes (84, 59, and 48 inches). They take about 30 minutes to put together and install onto barrels. The largest size collects 27 gallons of water per inch of rain. The water can be used for consumption, but should be purified first.

Keep in mind that you may be transporting the water from its source to your home. Look into backup solar pumps for your well, and be sure that a manual pump is also available. If water is going to have to be carried for any distance, consider what type of conveyance will make the job easier. As people age or become injured, the job of carrying two full buckets several times a day will become a lot more physically strenuous. A sturdy wheelbarrow, pushcart, or wagon would make the task easier.

SALTWATER

If you are really desperate and have no other options, can you drink saltwater?

Humans and animals CANNOT drink saline water without removing the salt first. So, if you live near an ocean or spend a lot of time near one, tread carefully before considering it a drinking water source.

Saltwater can be "desalinated" to make it safe for human consumption. In fact, it is a practice that is commonly used on ships, submarines, and in certain parts of the world, including Australia, Saudi Arabia, and Israel (which produces 40 percent of its domestic water through seawater desalination).

There are several ways to remove saline from water. We will discuss that, and how to purify water from even the most natural of sources, next.

CHAPTER NINE

TESTING YOUR WATER

We've already discussed the infinite possibilities for contaminants in water sources. Bacteria, viruses, parasites, nitrate, PPCPs, and toxic chemicals could be lurking beneath the surface of virtually any water source you can think of. It is safest to assume at least some of those pollutants and impurities are present and plan accordingly.

Even if you are getting presumably safe "city water" from a municipal supply, you should be provided with an annual report that explains what kind of testing was done on your water and what was found, if anything. Of course, if you aren't the trusting type, you can still test that water yourself as an added precaution.

If you have a well or are collecting water from a source that is not monitored and regulated, you will need to take responsibility for testing and purifying your water yourself.

Studies have shown that around 50 percent of private water systems fail at least one drinking water standard.[24]

Many common pollutants do not cause water to smell, taste, or look funny, so you can't rely on your senses to determine safety.

Water is a "universal solvent," meaning that it has the ability to dissolve almost anything it comes into contact with. This characteristic means that it is very easily contaminated.

Most testing isn't expensive, and the time and financial investment will provide you with priceless peace of mind. Not only is your family's health at stake, there are possible legal consequences involved. Think about how litigious our society is: If someone consumes your water and becomes ill, you'll want to be able to prove that you conducted the proper testing on a regular basis. And, should you suspect your water supply has become contaminated by an outside source, you'll want to have documentation to support your case.

---

24 http://extension.psu.edu/natural-resources/water/drinking-water/water-testing/testing/testing-your-drinking-water

TESTING KITS

You can test your water yourself or have a professional lab or service do it for you. Drinking water quality test kits are available for purchase online and at most superstores and home improvement stores. Basic kits usually test for bacteria, lead, nitrates/nitrites, pesticides, chlorine, hardness, and pH. They are fast, simple to use, and inexpensive: the price range for these kits is usually $10 to $40. Kits that test for less-common contaminants are also available. Some test for 15 or more contaminants, including the ones in the basic testing kits, plus iron, sulfate, copper, and sulfide.

Even more in-depth testing kits are available, but most of them require you to send your samples to a professional lab. Most of them check your water for around 100 different contaminants, including volatile organic compounds, toxic metals, heavy metals, and bacteria. The pricing for these comprehensive kits is typically in the $100 range, and results can take about a week to receive.

WHAT TO TEST FOR

At a bare minimum, you should test your water once a year for coliform bacteria and nitrates because of the serious health risks associated with those contaminants.

It is best to test for nitrate during the spring or summer following a rainy period, if possible.

If someone in your household becomes pregnant, test your water supply for nitrate in early months of the pregnancy. Test it again before bringing a newborn home, and again during the first six months of the baby's life. Remember, in the body, nitrate is converted into nitrite, which can cause brain damage and death in infants because it reduces the amount of oxygen in the baby's blood.

Test for total dissolved solids and pH every one to three years. These tests will provide you with an overall picture of the health of your water. The total dissolved solids content of drinking water should be below 500 mg/L. This value should not change much from test to test. If it does, further testing is necessary because it is likely that pollution has occurred.

LEAD

Lead is a naturally occurring element that can be found in air, soil, and water. Lead from natural sources is present in tap water to some extent, but analysis of both surface and groundwater suggests that lead concentration is generally fairly low. The main source of lead in drinking water is (old) lead piping and lead-combining solders. Homes that were built before 1986 are more likely to have pipes made of lead, but even "lead-free" piping can contain up to 8 percent lead. If you don't have lead pipes in your house, your water probably doesn't contain any; it is rarely found in source water.

Even though it is unlikely that your water supply contains lead (unless you have lead pipes), testing for it is a good idea.

Lead can damage various systems of the body, including the nervous and reproductive systems, the kidneys, and the bones. It also can cause high blood pressure and anemia and can interfere with the body's use of calcium and vitamin D. High amounts of lead in the blood of children can cause learning disabilities, behavioral problems, and mental retardation, all of which may be irreversible. At very high levels, lead can cause convulsions, coma, and death.[25]

If your water source tests positive for lead, you'll need to use a filtration system that is certified for lead removal or find a safer drinking water source.

ARSENIC

Something else you don't want in your water supply is arsenic. This naturally occurring element is found in rocks, soil, water, air, plants, and animals. Natural events like volcanic activity, forest fires, and erosion of rocks can cause it to be actively released into the environment. Arsenic is also used in agricultural and industrial practices and is used in some fertilizers, paints, dyes, metals, drugs, and soaps. It is also used as a wood preservative and can be released by mining and coal burning.

---

25 http://www.who.int/water_sanitation_health/diseases/lead/en

Arsenic is highly toxic and can affect nearly every organ system in the body.

There are short- and long-term health effects associated with arsenic exposure. Some effects appear within hours or days of exposure, and others develop over many years.

Long-term exposure to arsenic through drinking contaminated water can cause chronic arsenic poisoning, leading to life-long problems. This most commonly affects the skin in the form of lesions, discolorations, thickening, and cancer. Cancer of the bladder, lungs, prostate, kidneys, nasal passages, and liver are other possible devastating diseases arsenic can cause.

Arsenic can also affect the cardiovascular, pulmonary, immunological, neurological (with symptoms including numbness and partial paralysis), reproductive, and endocrine systems.

Severe arsenic poisoning can cause vomiting, abdominal pain, and diarrhea. These symptoms are followed by numbness and tingling of the extremities, muscle cramping, and in extreme cases, death.[26]

Water that contains high amounts of arsenic should not be used for drinking, cooking, or watering crops. Plants can take up arsenic through their roots, causing the product of the plant to contain high levels of arsenic, which is then passed on to the person or animal who consumes it. Rice has been found to have particularly high levels of arsenic, so much so that many

26 http://www.who.int/mediacentre/factsheets/fs372/en

holistic nutrition experts recommend eating rice infrequently or not at all.

Groundwater sources tend to have higher levels of arsenic than surface water sources. That's because the demand on groundwater is usually higher. It is more commonly used in municipal systems and private wells. This heavy use can cause water levels to drop, allowing arsenic to be released from rock formations.

Certain regions of the United States tend to have higher levels of arsenic in their water supplies. The EPA's standard is 10 parts per billion (ppb), and some western states have levels that are higher than that. Some parts of the Midwest and New England have levels that high, or close to it.[27]

Because of this toxic element's prevalence in the environment, testing your water source for arsenic contamination is a good idea. Most home-testing kits cost less than $15, and you'll see your results within minutes.

RADON

Radon is a gas that comes from the natural radioactive breakdown of uranium in the ground. It has no color, odor, or taste. Radon can dissolve and accumulate in groundwater, which means it can be found in water from wells. Not all ground-

27 http://water.epa.gov/lawsregs/rulesregs/sdwa/arsenic/Basic-Information.cfm

water contains radon, but drinking water that contains it can cause internal organ cancers like stomach cancer.

You can buy a simple kit to test your water source for radon, or you can contact your state radon office for assistance.

FLUORIDE

Fluoride is an ionic compound that contains a reactive element called fluorine. It is naturally found in many rocks.

Because it is believed to protect teeth from decay, it has been added to public water supplies since the 1940s. By 2010, water fluoridation had become widely used in the US, reaching about 204 million people.[28] This is also the main reason my family never, ever consumes municipal water if we are in an area that deliberately adds the compound to the public supply.

The incidence of tooth decay has declined in the United States since fluoridation began; however, it has also declined in other countries that do not fluoridate. Many argue the reduction in tooth decay is because of more accessible dental care and better dental hygiene, not water fluoridation.

Backing them up is research conducted within the last 15 years that has shown that fluoride primarily works topically, such as when it is applied to the teeth in toothpaste that contains fluoride.

Water fluoridation has been the subject of much controversy, and for good reason. Studies have shown that fluoride

28 http://www.cdc.gov/fluoridation/statistics/2010stats.htm

intake may cause a startling array of serious health problems, including increased risk of bone fractures, thyroid disorders, impaired immune system functioning, and cardiovascular disease. There is also some evidence that fluoride can cause osteosarcoma, a form of bone cancer. Researchers suspect a connection to cancer because half of ingested fluoride is deposited in bones, and fluoride stimulates growth in the end of bones, where osteosarcoma occurs.[29]

A study published in the fall of 2012 in *Environmental Health Perspectives* found a link between high fluoride levels found naturally in drinking water in China and elsewhere in the world, and lower IQs in children. The paper looked at the results of 27 different studies, 26 of which found a link between high-fluoride drinking water and lower IQ. The average IQ difference between high and low fluoride areas was 7 points, the study found.[30]

Children aged eight years and younger have an increased chance of developing dental fluorosis. In mild cases, this shows in white streaks on the teeth. In severe cases, it can include brown stains, pitting, and broken enamel. As of 2010, 41 percent of children from ages 12 to 15 had some level of dental fluorosis, according to the Centers for Disease Control and Prevention.[31]

---

29 http://healthydebates.com/15-facts-people-dont-know-fluoride

30 http://www.livescience.com/37123-fluoridation.html

31 http://www.cdc.gov/nchs/data/databriefs/db53.htm

Fluoride consumption over a lifetime may increase the likelihood of bone fractures, and may result in skeletal fluorosis, a painful and potentially crippling disease. The EPA has determined that safe exposure of fluoride is below 4 mg/L in drinking water to avoid those effects.

Naturally occurring fluoride concentrations in surface waters are generally low, but that depends on location. However, groundwater can contain much higher levels than the 4 mg/L recommended maximum.

Community water systems in areas with levels higher than that are required to lower the fluoride level below the acceptable standard. But the levels in private water sources, such as wells, may still be higher.

This means you will need to test your well water for fluoride and will need to remove the fluoride if your levels are above 4 mg/L.

WHEN SHOULD YOU TEST YOUR WATER?

Even if your water is crystal-clear, odorless, and tastes great, you still should test it for contaminants and pollutants on a regular basis. But sometimes there are signs that your water supply may need to be tested even more frequently. Here are some of those signs and what they might mean.

TASTE AND ODOR

- A strong chlorine taste or smell generally occurs when the water is treated at a water treatment plant to disinfect it and kill off bacteria and other harmful microorganisms.
- Some water systems have a high mineral concentration, resulting in a salty or soda-like taste. In the case of iron and manganese, a strong metallic taste is noticeable.
- A rotten egg smell is usually a result of decaying organic material underground. As water flows through these areas, hydrogen sulfide gas is picked up. When the water reaches the surface or comes out of your faucet, the gas is released into the air. Hydrogen sulfide gas is what produces the rotten egg smell. In large enough quantities, it is toxic to aquarium fish. You'll be able to taste as little as 0.5 (ppm) in your water.

  If your water smells like rotten eggs, it also may indicate the presence of bacteria.
- Musty or other unnatural or unusual smells are normally a result of organic matter or even some pesticides in the water supply. Even very low amounts can make your water smell funny.
- A turpentine taste or odor can be a result of MTBE contamination in your water. MTBE (methyl tertiary

butyl ether) is a flammable, colorless liquid fuel oxygenate chemical that dissolves easily in water. MTBE is added to gasoline to increase its oxygen content to lower carbon monoxide and other air pollutants that are emitted from vehicles.

While MTBE may help reduce air pollutants, it certainly isn't good for your drinking water. It spreads quickly through water and can easily contaminate it. This includes private drinking water systems like wells. Even a small amount will make your water undrinkable. According to the EPA, MTBE has not been used in significant quantities in gasoline since 2005. But groundwater in some areas of the US might still contain MTBE. It can enter water sources through leaking underground or aboveground gas storage tanks and pipelines, as well as from gasoline spills. It isn't known if MTBE causes health problems in humans, so it is best not to drink water that contains it.

COLORS

Your drinking water should be clear. Here is a list of possible coloration issues you may encounter, and what they may indicate.

- A red, brown, or rusty color is generally a sign of iron or manganese in your water. Iron in your water may cause stains in sinks or your laundry.

A bit more on iron and manganese: While these metallic elements may cause frustration if they stain your laundry or sinks, they generally are not harmful to health. But it is important to find out what type of iron is contaminating your water. That's because there are three kinds: ferrous iron, ferric iron, and iron bacteria. You'll want to treat your water to remove all three, but especially iron bacteria, because while they are not known to cause disease, they often help create an environment that is friendly to more harmful types of bacteria. Iron bacteria can also make your water taste and smell terrible. If you notice a cucumber or sewage-like smell coming from your water, the likely source is iron bacteria.

- A yellow color occurs in regions where the water has passed through marshlands and then moved through peat soils. In the United States, this is more likely to occur in the Southeast, Northwest, New England, and Great Lakes regions. It is more commonly found in surface water supplies and shallow wells. Although the yellow color may be displeasing, it presents no health hazard, as it is only small particles suspended in the water.

OTHER REASONS TO TEST YOUR WATER

- There is recurring gastrointestinal distress in your family or visiting guests.
- You are pregnant or have a child less than six months old living in your household.
- Your well is next to a septic tank, and it is questionable if the septic tank is placed far enough away from your well.
- Your property has an underground storage tank that is close to your well.
- Your property has a leaking gas tank that is next to your well.
- You have a new well and want to test the purity of your water.
- Your well is next to an area where livestock are kept.
- You have mixed pesticides or other chemicals near your well, or accidentally dropped these into your well.
- You have noticed an increased amount of turbidity (cloudiness) in your water.
- Your property is near a chemical plant, gas station (either abandoned or not), mining operation, landfill or dump, dry cleaner, junkyard, heavily salted roadway, or oil or gas drilling company.

- A green or blue color generally indicates that there is copper in your water supply, or copper pipes and corrosive water. The copper can cause staining of your fixtures and your laundry. Copper is regulated in drinking water by the EPA at 1.3 ppm. This is at a low enough concentration that the copper won't be tasted (the taste threshold is around 5 ppm). However, copper can become a problem if it is higher than 30 ppm in your water. At this level, copper can cause vomiting, diarrhea, and general gastrointestinal issues. If you are using well water as your primary source of water and copper is a concern in your area, it would be to your advantage to have your water tested for copper.

- Cloudy, white, or foamy water is usually due to turbidity. Turbidity is caused by finely divided particles in the water. When light hits the water, it is scattered, giving a cloudy look to the water. The particles may be of either organic or inorganic in nature. Cloudiness itself isn't dangerous, but the cause of it may be.

CHAPTER TEN

PURIFYING WATER

There are many different ways to purify water. The method or methods you choose will depend on your water source, how much water you need to purify, and what kind of contaminants you may be dealing with. In most cases, you'll want to combine these two steps: purification and filtration.

Purification removes microscopic things like bacteria and viruses, while filtration removes the "chunks" from your water, like silt and natural debris.

During a crisis situation, you want to be particularly careful. You've already got a disaster on your hands, and medical assistance could be hard to come by. Don't make a bad situation even worse by taking chances with your water.

BOILING

Boiling water will kill most pathogens, but will NOT remove chemicals from water. In fact, boiling water can increase concentrations of things like nitrate in water because they increase as water evaporates. If your water has been contaminated by sewage, boiling will not be enough to purify it.

Boiling DOES help kill certain pathogens, though. The standard recommendation is a full rolling boil for one minute. Let the water cool before using it. At elevations higher than 6,500 feet, boiling time should be extended to three minutes. Boiling water by those guidelines should kill:

- Protozoa (including cryptosporidium and Giardia intestinalis, also known as Giardia lamblia)
- Bacteria (including campylobacter, salmonella, shigella, and E. coli)
- Viruses (including enterovirus, hepatitis A, norovirus, and rotavirus)

Boiled water will taste better if you reoxygenate it by pouring the water back and forth between two clean containers. Also, be careful not to boil your water for too long; the longer you do, the more of it will evaporate.

Boiling may not always be practical, though, so read on for other methods to remove the above pathogens.

CHEMICAL TREATMENT

Chlorine and iodine are the two chemicals most commonly used to treat water. Both are easy to use, portable, and low-cost, but need to be used with caution and according to specific instructions.

For chemical disinfection to be effective, the water must be filtered and settled first. Chlorine is generally more effective than iodine in controlling giardia, and both disinfectants work much better in warm water.

Treating water with iodine or chlorine can kill bacteria and viruses, but not all protozoan cysts. For example, cryptosporidium, a microscopic parasite that causes the diarrheal disease cryptosporidiosis (commonly known as "crypto"), shows strong resistance to iodine and chlorine, so you'll need to boil or filter water that may be infected with this parasite before drinking it.

In fact, if you have the option, combining chemical treatment with filtration is better than chemical treatment alone.

IODINE

Iodine comes in three forms: tablets, tinctures, and crystals.

There are important things to know about using iodine to treat water. Iodine can be harmful, especially to children, pregnant women, and people with thyroid conditions. Some people are allergic to iodine—those who are allergic to shellfish are especially at risk for an iodine allergy.

Iodine-treated water should NOT be used by anyone for periods longer than 14 days. Very careful attention must be paid to using the correct measurements and instructions on the manufacturer's label. Typically, the colder the water, the longer it will take for iodine to work and the less effective the treatment may be.

Iodized water can impart a taste that some people don't like. This can be remedied by adding vitamin C in powder or tablet form. You'll need to wait at least 30 minutes after treating water with iodine to add vitamin C, though, because that's how long the iodine needs to do its job. Kits are available that include an iodine pill and a vitamin C pill. You can also keep some powdered drink mixes on hand, like lemonade or Tang, to disguise the flavor of your water.

Commercially prepared iodine tablets containing the necessary dosage for drinking water disinfection are available at drugstores. These are good to have on hand in case of an emergency and are simple to use. Follow the manufacturer's instructions on the package. If directions are not available, use one tablet for each quart or liter of filtered and settled water to be purified.

You can also use a tincture of iodine to disinfect filtered and settled water. Common household iodine from the pharmacy, your medicine chest, or first aid kit may be used. Add five drops of 2 percent tincture of iodine to each quart or liter of clear water. For cloudy water, add 10 drops and let the solution stand for at least 30 minutes.

Another product to keep on hand is iodine crystals. Polar Pure is perhaps the most popular brand of iodine crystals available. Here's how you use this product:

Fill the Polar Pure bottle with water and shake it. The solution will be ready for use in one hour. Add the number of capfuls (per quart of water treated) listed on the bottle to your container of water based on the temperature of the iodine solution. The particle trap in the bottle prevents crystals from getting into the water you are treating. It is important to note that you are using the iodine solution to treat the water, not the iodine crystals. The concentration of iodine in a crystal is poisonous and can burn your skin and eyes. Let the treated water stand for 30 minutes before drinking. In order to destroy giardia cysts, the drinking water must be at least 68 degrees. If you are outside, your water can be warmed in the sun before you treat it, or hot water can be added. Refill the treatment bottle after use so that the solution will be ready one hour later. The crystals in the bottle make enough solution to treat about 2,000 quarts. Keep this out of the reach of children, as the crystals are toxic.

CHLORINE

You can use unscented, household chlorine bleach that contains a chlorine compound to disinfect water. Did I mention it is vital to use unscented bleach? You do NOT want perfumes in your drinking water. Your water can be filtered before or after treating it with chlorine.

Do not use non-chlorine bleach to disinfect water. Typically, household chlorine bleaches will be 5.25 percent available chlorine. Follow the procedure written on the label.

When the necessary procedure is not given, find the percentage of available chlorine on the label and follow these guidelines:

DISINFECTING WATER WITH CHLORINE

% of Available Chlorine	Drops per Quart	Drops per Gallon	Drops per Liter
1%	10	40	10
4% to 6%	2	8 (⅛ teaspoon)	2
7% to 10%	1	4	1

Mix the treated water thoroughly and allow it to stand (ideally, covered) for 30 minutes. The water should have a slight chlorine odor. If it doesn't, repeat the dosage and allow the water to stand for another 15 minutes. If the treated water has too strong of a chlorine taste, allow the water to stand exposed to the air for a few hours or pour it from one clean container to another several times.

You also can use granular calcium hypochlorite (also known as "pool shock") to disinfect water. Unlike regular household bleach, it doesn't degrade or expire. Here's how to use it:

- First, you'll have to make a stock of chlorine solution. This is NOT for drinking. You will be adding drops of this solution to much larger volumes of water as you need it.
- To make the solution, dissolve 1 teaspoon (about ¼ ounce) of high-test (78 percent) granular calcium hypochlorite for each 2 gallons of water, or 5 milliliters (approximately 7 grams) per 7.5 liters of water.
- To disinfect water, add one part of this chlorine solution to 100 parts water to be treated. Let the mixture sit for at least 30 minutes before drinking.
- If your treated water has a chlorine odor that disturbs you, you can aerate the disinfected water by pouring it back and forth from one clean container to another.

You can use chlorine tablets like Potable Aqua to disinfect filtered and settled water. These are portable and easy to carry with you, but they take longer to work than other disinfection methods. Most of these kinds of tablets require a four-hour treatment time.

No matter what form of chlorine you decide to use, please remember to use it with caution and keep it out of the reach of children. Chlorine is very poisonous when it isn't used properly.

FILTRATION

Simply put, filtration means to strain out impurities from water. Purifiers do the same, but also can remove viruses.

Simple water filters, mechanical filters, activated carbon filters, oxidizing filters, neutralizing filters, and microfilters fall into this category.

The simplest way to create an emergency water filter is to run the water through a cloth. The tighter the weave of the cloth, the better, as it will allow fewer particles to flow through. You also can fold the cloth into multiple layers to make it more likely to catch particles. If you happen to be away from your supplies, you can always fall back on a clean bandana to filter your drinking water before purifying it.

An important addition to your water preparedness supplies is a high-quality filtration system.

Most filtering elements are made of ceramic, glass fiber, hard-block carbon, or materials that resemble compressed surgical paper.

One of the most commonly used filtration methods is activated carbon filters. They are popular because they not only improve the taste and odor of water, they can effectively remove many chemicals and gases, including residual chlorine, some metals like mercury, many organic compounds (like VOCs), some pesticides, gasoline, trihalomethanes (THMs, the chlorine byproduct), benzene, and radon gas.

But there are drawbacks. Activated carbon filters (AC filters) generally will not remove total dissolved solids or most heavy metals. In fact, only a few carbon filter systems have been certified for the removal of lead, asbestos, cysts, and coliform, all of which can be present in water and pose significant health risks.

A special type of AC filter called a solid block activated carbon filter can remove cryptosporidium and giardia cysts. However, AC filters cannot remove bacteria, nitrate, fluoride, and chloride, or improve water hardness (caused by calcium and magnesium).

The pores of AC filters trap microscopic particles and large organic molecules, while the activated surface areas cling to (adsorb) small organic molecules.

The ability of an activated carbon filter to remove certain microorganisms and certain organic chemicals, especially pesticides, trihalomethanes, trichloroethylene (TCE), and PCBs depends upon several factors, such as the type of carbon and the amount used, the design of the filter and the rate of water flow, how long the filter has been in use, and the types of impurities the filter has previously removed.

If activated carbon filters aren't replaced as recommended, they can become more harmful than using nothing at all. That's because the filters themselves can harbor bacteria, which can contaminate your water.

Because of their limitations, activated carbon filters are best used in conjunction with another purification method, like reverse osmosis.

REVERSE OSMOSIS

Reverse osmosis (RO) is a membrane separation process that uses a very thin membrane with tiny pores to separate out most of the dissolved minerals and suspended particles from water. The membrane is semipermeable, meaning that it allows pure water to pass through but retains any dissolved and suspended constituents.

In other words, reverse osmosis forces water through a special membrane that allows water molecules through, but stops larger molecules like lead, chromium, and arsenic from getting by.

In fact reverse osmosis is very good at removing many contaminants from water. This includes total dissolved solids, asbestos, nitrates, radium, pesticides, chlorinated particles, copper, fluoride, and VOCs. RO is also effective in eliminating bacteria, viruses, Giardia lamblia, and cryptosporidium.

Combining reverse osmosis with activated carbon filtration will help you remove most impurities and contaminants from your drinking water supply, and is generally considered the best overall water purification method.

There are downsides to using reverse osmosis units: they use about three times as much water as they treat, and they work slowly.

DISTILLATION

Distillers heat water to the boiling point and then collect the water vapor as it condenses. The process kills most disease-causing microbes and eliminates most chemical contaminants. Some contaminants, like those that easily turn into gases (such as gasoline components or radon) may remain in the water unless the system is specifically designed to remove them. To some people, distilled water may taste flat because the water's natural minerals and dissolved oxygen often have been removed (this issue can occur with reverse osmosis systems as well).

There are a few significant drawbacks to distillation. Herbicides and pesticides have boiling points lower than 212°F (100°C), so they not only cannot be removed efficiently through distillation, they can actually become more concentrated in your water. And distillation isn't cheap: the process requires large amounts of energy and water. The distillation process is slow, too. It takes about four to five hours for the typical home unit to produce about a gallon of water.

ULTRAVIOLET (UV) LIGHT

Ultraviolet light has been used to disinfect public water supplies for over 75 years.[32] Sunlight's germicidal properties were discovered in 1877. After it was determined that UV wave-

32 http://viqua.com/why-uv

lengths were responsible for the germicidal activity, researchers began implementing its use in water purification systems.

UV light destroys disease-causing microorganisms by penetrating them and attacking their genetic core, or DNA. This process is called thymine dimerization. It can inactivate bacteria, viruses, and protozoa such as cryptosporidium and giardia, preventing them from being able to reproduce or infect.

There are pros and cons to using UV light to purify water.

PROS

- UV purification does not use any chemicals like chlorine or leave any harmful byproducts. It does not add any chemical tastes or odors to your water.
- It is one of the most effective ways to kill disease-causing microbes.
- UV units require very little energy—they use about the same amount of energy as it takes to run a 60-watt lightbulb.
- The units are low maintenance: once you set up your system, you'll just need to change the UV bulb every year.

CONS

- UV light on its own is not enough to purify water for drinking. That's because UV light does not eliminate contaminants like chlorine, heavy metals, and VOCs.

- UV systems are often used in conjunction with reverse osmosis systems to provide a complete purification process for safe drinking water.

GRAVITY FILTERS

Gravity filters do not require power to work. Some gravity products, like Berkey water filters,[33] are unique in that they both filter and purify water. This is the product we use because it's the most thorough off-grid method of making your water safe to drink. Berkey products use an ionic adsorption process combined with microfiltration, which creates a pore structure so minute that contaminants cannot pass through the charged filter.

They can remove viruses, bacteria, and residual chlorine, and reduce heavy metals, VOCs, arsenic, and nitrites.

In addition to products you can use in your home, companies like Berkey, Platypus, and Sawyer offer gravity filtration systems that are portable and can be taken with you.

My favorite "to-go" system is the Sawyer Mini. It's a tiny little pump system that fits easily in a backpack or purse and can be used to cleanse hundreds of gallons of water. This should definitely be part of your EDC (everyday carry) kit and part of your BOB (bug out bag).

---

33 http://www.berkeyfilters.com/berkey-answers/performance/how-berkey-purifies

CHAPTER ELEVEN

FINDING EMERGENCY SOURCES OF WATER

There may come a time when, for one reason or another, you do not have enough water stored for an emergency and backup resources are unavailable.

In these extreme situations, you'll need to get creative and find alternate sources of water. There are resources inside and outside of your home that you should be able to tap into if the unthinkable happens.

Please use your common sense. DO NOT drink water that has an unusual odor or color. Obviously, avoid water that may have been contaminated with fuel or toxic chemicals.

If your home's water service is interrupted due to a natural disaster in your area, listen to reports from local officials for advice on water precautions in your home. You might need to take the added step of shutting off the main water valve to your house to prevent contaminants from entering your pipe system.

The following are possible sources of water.

INSIDE OF YOUR HOME

You don't necessarily have to leave your home (and in some cases, may not be able to) to find hidden water sources.

FREEZER

The first surprisingly obvious place to look would be your freezer. Do you have ice cube trays or an automatic ice cube maker? Chances are, you do, and hopefully, you have an abundance of ice cubes that you can melt and use for drinking water. Place your ice cubes in containers with lids and allow them to melt at room temperature.

If you have cans of fruits and vegetables, the liquid from those can be used to drink as well.

WATER HEATER TANK

The average home water heater holds 30 to 60 gallons of water. This is a very valuable source of water that you can use in times of desperation.

To obtain water from your hot water tank, follow these instructions:

Note: Use caution when draining water from a hot water heater tank. If the water heater was working prior to the disruption, the water will typically be 120°F/49°C to 140°F/60°C. Don't burn yourself!

- Turn off the gas or electrical supply to the tank.
- Close the water intake valve into the tank by closing the shut-off valve at the top of the tank. This helps preserve the quality of the water in the tank.
- Turn on a hot water faucet in your home to allow air into the tank, which will allow water to drain.
- Open the water heater drain faucet briefly to rinse the interior surfaces and then catch the water in a container.
- DO NOT turn the gas or electricity back on until the water supply is restored and the tank is full of water.
- After the water has cooled, treat it by adding six drops of liquid household bleach for every gallon of water. The bleach should contain 5.25 percent sodium hypochlorite. Stir in the bleach and let the water stand

for 30 minutes. If, after 30 minutes, the water doesn't smell like chlorine, add another 6 drops of bleach and let it stand for 15 minutes.

If you have a good-quality water filtration and purification system, you can use that to help ensure the water from your heater is safe to drink.

This is a good time to remind you that most professionals recommend flushing and draining your hot water heater annually. Doing so will not only improve its functioning, but will help reduce the amount of sediment that builds up, providing you with a huge supply of water should an emergency occur.

PIPES

Draw the water out of your pipes. Turn on the faucet at the highest level in your home (for example, the upstairs bathroom). A small amount of water will trickle out and then air can flow into the pipes.

Then, turn on a faucet at the lowest level. If there is a faucet in your basement, use that one. If not, use one in a lower-level bathroom or your kitchen sink. Drain the water into storage containers.

TOILET

If you are really desperate, the water in the flush tanks of your toilets can be used if necessary. I would only use this water as a last resort and ONLY if I were absolutely positive it was NOT treated with chemicals. This seems like a no-brainer but I'll say it anyway: do NOT use the water in the BOWL!

Scoop the water from your flush tank out using a clean ladle or cup. Boil the water for three to five minutes or add a couple of drops of bleach to each gallon to purify it. Be sure to let water sit for 30 minutes before drinking it.

OUTSIDE OF YOUR HOME

If you have a swimming pool, it can provide a tremendous amount of water that can be used in an emergency situation. There are some caveats here, of course. If your pool is poorly maintained and has algae growing in it, it is going to be trickier to treat properly to make it safe for consumption. As we discussed earlier, your swimming pool water is only really viable for the first day or two after an emergency, because once the pump stops working, the algae growth will be abundant.

Should a swimming pool be your only option, if possible, test the chlorine level to see what you are dealing with. Remember, the EPA says that water with chlorine levels of up to 4 ppm is generally considered safe to drink. You'll want to boil it and use whatever filtration and purification system you have to further decontaminate the water, too.

STREAMS, RIVERS, CREEKS, AND OTHER MOVING BODIES OF WATER

Moving water is typically cleaner than stagnant water because it has been washing over rocks and other natural forms of purifiers. At the very least, you will still have to boil it, especially after a rain. That's because when it rains, animal waste is more likely to find its way into moving streams.

RAINWATER

If you own a device like RainSaucers, great! They can really come in handy in emergency situations—IF there is rainfall. Placing clean buckets and pails outside to collect rainwater will work, too, but be sure you don't use containers with chemical residues. Ideally, you'll have some clean buckets around just in case you need them. While you can throw together a system in an emergency, it's far better to have your harvesting setup created in advance.

PONDS, LAKES, SPRINGS, AND OTHER BODIES OF WATER

If you live close to one of these bodies of water or are able to travel to one, they can be tapped into for water if necessary. You'll need to be sure to follow filtration and purification methods to prepare the water for consumption, of course. For saltwater, use a purification method that will desalinate it (like reverse osmosis or distillation) prior to drinking. Saltwater is very dangerous to consume without removing the saline. It will dehydrate you, which defeats the purpose of drinking it in the first place. Drinking it in large amounts can cause kidney damage.

CREATE A SIMPLE SOLAR STILL

If you have sunlight, you can harness its power to draw water from the earth. You'll need a clean container to catch water, tarp or plastic sheet, and bricks or large rocks. Here's how to do it.

- Dig a hole in the ground in an area that gets a lot of direct sun. The size of the hole will depend on how quickly you wish to extract water, how large your tarp or plastic sheet is, and how much energy you have for digging.
- After you dig your hole, place a container in the center.
- Lay the tarp or plastic sheet across the diameter of the hole. Use bricks or rocks to seal the perimeter.
- Place a rock in the center of the tarp, just heavy enough so that the center of the tarp slopes downward. Don't poke any holes in your tarp or plastic sheet. Water will be generated from condensation on the underside of the tarp. The moisture generated from the heat of the sun beating down on your tarp

will drip down toward the center and into your container.

The water you obtain by this method will not need to be purified, as it is already naturally clean. It can take a long time to generate water this way, but it is an alternative you can add to your options.

CHAPTER TWELVE

SANITATION

More people die after a natural disaster than during it, and it's usually from waterborne illnesses that occurred because of substandard sanitation practices. It only takes ONE person who deals with waste incorrectly to contaminate the water supply and sicken thousands.

It is crucial to know the best ways to handle washing, cleaning, and dealing with human waste so water supplies are not contaminated.

When Hurricane Katrina rampaged through New Orleans on August 29, 2005, she brought 100 to 140 mph winds and left massive destruction. But the aftermath of Katrina's fury was far, far worse than the storm itself. The most obvious risk associated with floodwater is drowning. But what lurks in those waters can kill you, too.

After Katrina, officials were concerned that the prolonged flooding would lead to an outbreak of health problems for people who remained in the city. Dehydration and foodborne illnesses were serious potential dangers, but they weren't the only hazards residents faced.

Flooding, combined with the city's heat and stifling humidity, increased the risk of the spread of hepatitis A, cholera, tuberculosis, and typhoid fever. People could also face long-term health risks due to prolonged exposure to the petrochemical-tainted floodwater and mosquito-borne diseases such as yellow fever, malaria, and West Nile Virus.

The storm's floodwaters covered nearly 80 percent of the city. When the water finally receded, it left behind a thick layer of sediment that coated streets, yards, and playgrounds. The Environmental Protection Agency, state agencies, and researchers collected soil samples from around New Orleans. Their findings were disturbing: arsenic and lead levels in many areas far exceeded state and federal standards designed to protect health.

Heavy rains and flooding can contaminate drinking water with sewage, petroleum products, pesticides, herbicides, and waste from farm animals. Floodwaters may contain more than 100 types of disease-causing bacteria, viruses, and parasites.

Floods also result in water damage, which increases the risk of toxic mold growth in homes. Lakes, rivers, and ponds can become contaminated, and sewage can back up into yards and basements if local sewage lines and septic tanks overflow.

For weeks or months after a natural disaster, health risks can continue to plague impacted areas. The standing water that remains after a flood can be a veritable cesspool of disease. Avoid it if you can, and if you cannot, be sure to follow good hygiene practices afterward. Do not allow children to play in floodwaters, and do not let them play with toys that came into contact with floodwaters and have not been disinfected.

PERSONAL HYGIENE IN AN OFF-GRID SITUATION

One of the most important things you can do to stay healthy in the aftermath of a disaster is to simply stay clean. Some people think that this is the least of their worries, but good personal hygiene greatly reduces the risk of you and your family members becoming ill. Many infections and waterborne illnesses are the result of poor hygiene. The dirtier and germier you are, the more likely it is that bacteria and viruses will be introduced to your system, making you sick.

Then there's the mental aspect. If you are generally a minimum-one-shower-per-day person with squeaky clean hair and perpetually fresh-smelling armpits, you're going to be very bothered by the unpleasant reality that your hair is greasy, your face is grimy, and quite frankly, you stink. You'll feel much more confident when dealing with FEMA or the insurance adjuster or life in general if you are relatively clean and well-groomed.

Of course, when the water isn't flowing from the taps, it's much more difficult to stay clean. Following, you'll find some practical solutions to help with off-grid hygiene.

Your first line of defense after any natural disaster or exposure to infectious materials is handwashing.

HANDWASHING 101

Keeping your hands clean during an emergency helps prevent the spread of germs. If your water supply is not safe to use, boil or disinfect prior to using it for handwashing.

Follow these steps for proper handwashing:

1. Wet your hands with clean, running water (warm or cold), turn off the tap, and apply soap.
2. Lather your hands by rubbing them together with the soap. Be sure to lather the backs of your hands, between your fingers, and under your nails.
3. Scrub your hands for at least 20 seconds.
4. Rinse your hands well under clean, running water.
5. Dry your hands using a clean towel or air dry them.

If your water service is interrupted or your tap water is contaminated, you can create a handwashing station. A portable water cooler with a spigot is ideal for this. Disinfect the cooler or large jug that you'll be using for your sanitized water. Fill the cooler or jug(s) with clean water (ideally, you'll use water that is clean enough to drink). Place a bucket or pail under the spigot to catch the water. Use paper towels to turn the water on and off; don't touch the spigot with your hands.

If you do not have access to water or want a second line of defense to use after washing hands, use an alcohol-based hand sanitizer that contains at least 60 percent alcohol. Alcohol-based hand sanitizers can reduce the number of microbes on your hands in some situations, but they are NOT effective against many germs. Do not depend on them as your primary hand-cleaning method. If your hands are visibly dirty, sanitizers won't do much to clean them. (Note: In normal circumstances, I don't recommend frequent use of hand sanitizer. In a crisis situation, however, hand sanitizer can help you deal with the added risk of illness.)

BABY WIPES

Baby wipes aren't just for babies. I've stocked up on a huge amount of baby wipe refills because in a down-grid situation, they're hard to beat for personal hygiene. It's really important to try them out on your skin before stockpiling a particular brand, though, because you don't want to end up with a bunch of wipes that have an ingredient to which you are intolerant. The apocalypse will be bad enough without an itchy rash in unmentionable places.

Here are a few ways you can use baby wipes in a down-grid scenario:

- **Wipe with these instead of toilet paper**. First of all, you can store a lot more baby wipes in a smaller amount of space than toilet paper takes up. Second, you'll get far cleaner, something that is especially important when you can't just hop in the shower every day.
- **Wash your hands**. You'll want to be especially careful to practice good hand hygiene during a disaster. When water is in short supply, baby wipes can provide a quick, easy way to wash your hands after using the bathroom, after handling something that could be contaminated, or before handling food.
- **Wipe off counters**. In a pinch, you can use baby wipes to clean surfaces. (I personally keep kitchen wipes with bleach for this purpose.) You can also use them to wipe down things like doorknobs, cupboard handles, remotes, and phones.
- **Bathe with them**. Okay, it's not ideal, but it's better than allowing yourself to become increasingly filthy. If you have no other option, you can give yourself or your child a "sponge bath" with baby wipes. More information on sponge baths follows.

SPONGE BATH 101

There's actually an art to taking a proper sponge bath. While it seems like it would be very simple to draw a basin of water and wash yourself, there are some tricks that make it a whole lot easier. When we lived at the cabin, I read a *Mother Earth News* article by Ole Wik, a man who lived off-grid in the Alaskan wilderness, that explained what I feel is the very best method for bathing without running water.[34]

Gather your supplies:

- Basin
- Hot water, if possible
- Baking soda
- Two washcloths

Do you see what's missing in that list? Soap.

The key to a successful sponge bath that doesn't leave you with itchy soap residue is, well, not using soap.

Before you say, "ewwwwww," remember that baking soda is a mild alkali. It will leave you feeling fresh and clean. In his article for *Mother Earth News*, Wik explains why this works so successfully:[35]

> *You see, all soaps are made by combining a fat and an alkali (usually lye). Baking soda—itself a mild alkali—seems to react with hair oils to produce its own natural, mild washing product. Under the proper conditions, soda will even create a copious lather.*

---

34 http://www.motherearthnews.com/nature-and-environment/sponge-bath-zmaz81mazraw.aspx

35 http://www.motherearthnews.com/nature-and-environment/sponge-bath-zmaz81mazraw.aspx?PageId=2#ArticleContent

When you "draw" your bath water, add a heaping teaspoon of soda to that liquid, too. Baking soda is a good cleaning and deodorizing agent, and I believe it has a beneficial effect on any kind of skin. (Pregnant women sometimes use it to relieve the itching sensation caused by their bellies' stretching.) My guess is that the mild alkali combines with skin oil—just as it does with hair oil—to form a natural soap. One thing's for sure: a soda wash leaves you feeling clean and refreshed.

So don't use soap. It will be very difficult to rinse off your body using just a basin, and you'll be itchy and uncomfortable. Obviously, do a little patch test before putting this into practice; you want to be sure that the baking soda itself doesn't bother your skin.

Now, back to the sponge bath. This procedure uses two basins of water. The basins can be of the type you'd use to wash dishes.

If it's cold in your house, set up some kind of privacy method and bathe as close to the heat source as possible. This has the added benefit of keeping you conveniently located to the next batch of hot water, should you need it.

Fill your basin with hot water and put in a big spoon of baking soda. Swish it around with your washcloth to dissolve it. Making sure the water isn't hot enough to burn you, dunk your head to get your hair wet.

Dump some baking soda in your hands and scrub your head and hair with it, paying special attention to the roots. If possible, use a wide-tooth comb to move the paste through your hair. Leave this on your head, and move on to your body.

Wet your washcloth and use it to dampen your body. Work in sections. Once you have a section dampened, place some baking soda in your hand and mix a little water with it to make a paste. Scrub the damp section of your body with the paste. Don't rinse yet. Cover your whole body with baking soda paste.

Now, take your washcloth and rinse it out well in the basin. Using a wet cloth, scrub any remaining paste off of your body, rinsing the cloth between passes. Draw a new basin of water and get the second washcloth. Give yourself one more scrub to make sure you get all of the baking soda residue off of your body.

Now, rinse your hair. Dunk your head in the basin (again, making sure not to burn yourself.) Scrub well to get the baking soda out of your hair.

Dry off, put on fresh clothes, and face the world, clean and confident.

HOW TO WASH CLOTHING BY HAND

If you've never washed a load of clothing by hand, it can seem like a daunting task. We had to do this up at the cabin for about half of the time we lived there, until we finally got a washing machine. First, gather your supplies:

- Laundry soap of choice
- Borax
- Baking soda
- Hydrogen peroxide
- Sturdy scrub brush
- Small bucket (I use a clean plastic kitty litter bucket)
- Plunger for agitating
- Good-quality janitor's mop bucket with a press wringer. DO NOT go cheap with this. You'll thank me later.
- Drying rack and clothespins (or drying method of choice)

Once your supplies are gathered, you're ready to wash. Here are a few tips to make it easier:

- **Save your shower water**. I know, lots of people are saying, "Ewww." But unless you're out working in a field or doing greasy mechanical work, you're probably not that dirty. Simply put the plug in the bathtub during your shower to collect the water. You can also put some gentle laundry soap in the bottom of the tub so you are ankle deep in bubbles.
- **Use your feet**. You can take it one "step" further. (Sorry, I had to make that pun.) Put your dirty clothes

in the tub and use your feet to "agitate" them while you shower.

- **Let it soak**. You can let the laundry soak in your tub full of shower water anywhere from half an hour to overnight. Unless your clothes are very, very dirty, you won't have much scrubbing to do after this.
- **Scrub the laundry**. Use a scrub brush and a combination of laundry soap and baking soda. This is where an old-fashioned washboard would come in handy, but if you don't have one, just use the bottom of the bath tub. Pay special attention to "dirty" areas: around collars, underarms, and knees, and soiled kitchen linens, socks, and undergarments.
- **Don't wear white socks**. I've learned that no matter how hard I scrub, nothing short of a tub full of bleach water gets our white socks looking clean, even though in terms of "sanitation" they are very clean. Investing in black socks for the stockpile will save you a lot of work should a long-term electrical disaster ever take place.
- **Some items require more soaking**. I use homemade "oxygen cleaner" in a treatment bucket for this: ⅛ cup each of baking soda, hydrogen peroxide, and laundry soap and approximately a half gallon of hot water.

- **Drain and refill your laundry tub**. Some people add more laundry soap here but I feel like the soap on the clothing from scrubbing it is sufficient. I have a broom handle that I use for stirring the laundry around, but there are items that look similar to a toilet plunger designed specifically for the purpose of agitating laundry. After agitating, allow it to soak again. I usually leave it for a couple of hours while I do other things. After the first hour, dump the items soaking in the treatment bucket into the big tub, along with the liquid in the bucket, and give it another stir.
- **Rinse your laundry**. Rinse your treatment bucket and drain the tub. Gently squeeze out the clothing items and let that water run down the drain or into a catchment system for flushing water. This will get some of the soap out. Add clean water to the tub, just enough to cover the laundry again. Rinse each item by swishing it vigorously through the water, then place it in the bucket that you soaked items in.
- **Wring it out**. Place your rinsed laundry into the wringer section of your mop bucket. Squeeze it out. Change the position of the item and squeeze it out again. This doesn't get your clothing wrung out as well as the spin cycle on the washing machine, but it's far more effective than wringing it out by hand. (Not to mention way easier on your hands!) You can

add this water to your collection of toilet-flushing water. Invest in the best quality-bucket you can afford. I bought a cheap one first and it broke after half a dozen loads of laundry. Consider this a tool that will take a beating. My bucket is an industrial-quality janitor's bucket.

- **It's still going to drip**. No matter how well you wring out your laundry, it's still going to drip for hours. I learned a little tip from my friend, Lizzie Bennett, who authors the British website, Under groundMedic.com: Place your drying rack in the bathtub for a few hours! This keeps your floors dry and keeps your home from becoming excessively moist. In the UK, few people have driers, so most air dry their clothing indoors in the bad weather.
- **Move it to a better place to dry**. I usually let the laundry drip in the tub overnight. After that, I relocate it outside, weather permitting, or by the woodstove.

This is a good skill to learn now, because in a down-grid situation, when water could be limited for a multitude of reasons, you don't want to waste your supplies and still have laundry that isn't very clean.

I added a really awesome product to my off-grid arsenal called a Wonderwash. It's a tabletop device powered by a hand crank. You can fit four to six T-shirts in it, and it saves a ton of work when you do your laundry manually.

The mental aspect of being able to don fresh, clean clothing in the aftermath of a disaster cannot be underestimated. Clean clothing is a sign of normalcy, and even more importantly, good hygiene will help prevent the spread of disease.

WASTE MANAGEMENT

So far, we've discussed how to handle personal hygiene after a natural disaster when your access to running water is interrupted. But if you do not have running water, there's another, albeit more unpleasant, issue you'll have to contend with: waste management.

If you can't flush your toilets, you will need to find alternate methods for your family's bathroom needs.

Bacteria, parasites, viruses, and other dangerous pathogens are abundant in human waste. In order to protect your family and community from illness, understanding how to properly handle waste is critical.

TO FLUSH OR NOT TO FLUSH

If you are on a septic system with no risk of the toilet backing up into the house, simply store some water for flushing in the bathroom. You won't want to waste your precious, clean drinking water supply on, well, flushing waste. At the first sign of a storm, we always fill the bathtub for this purpose. Add the stored tub water (or other water you can find that is not suitable for drinking) to the tank so that you can flush.

You also can flush your toilet with a bucket of water. This requires a full gallon of water, which you will pour directly into the toilet bowl.

Start slowly at first, then quickly pour the rest of the water into the bowl. The shape of the toilet and the pressure from the water in the bucket will push everything through the pipes, so you won't need to use the handle to flush.

Be sure you have an abundant water supply stored or have access to water you can use for flushing before you use one of the methods above. You don't want to run out of water because you've been literally flushing it all away.

If you are NOT on a septic system, you will NOT be able to flush your toilets at all as long as the water supply is shut off. To do so using the method above is to risk your sewage overflowing all over your bathroom floor.

This means you'll need an alternative.

TOILET SUBSTITUTES

One solution is to stock up on extremely heavy-duty garbage bags (the kind that contractors use at construction sites) and kitty litter. Place a bag in your drained toilet. Sprinkle some kitty litter in the bottom of the bag. Each time someone uses the bathroom, add another handful of litter. Be very careful that the bag doesn't get too heavy for you to handle it. Tie it up very securely and store it outside (out of direct sunlight) until services are restored. Label the bag "human waste."

Fireplace ashes or sawdust will work in place of kitty litter if necessary, but kitty litter is the best option because it has stronger deodorizing properties.

Another alternative is to make a portable potty. A 5-gallon bucket will work well. Just line it the same way as described for your toilet. If you don't have a 5-gallon bucket, any large pail or small trash can with a lid will work. To make it a bit more comfortable to sit on, you can remove your toilet seat and place it on the bucket, you can purchase a toilet seat made specifically for 5-gallon buckets, or you can make a seat with two 2x4s placed parallel across the bucket. Line it with heavy-duty garbage bags and sprinkle kitty litter in it like you would if you were using your regular toilet.

You will need to check with your local health department to find out how to dispose of the bags of waste. In most regions, simply tossing them into your garbage is not permitted.

OUTDOOR TOILETS

If you can't use your toilets and don't have any large buckets or receptacles you can use to make an emergency potty, you may have to resort to digging a simple latrine somewhere on your property. Find an area on your property that is at least 50 feet away from all water sources like wells, streams, lakes, or springs. This is crucial—you do NOT want human fecal matter reaching your water supply and making people and animals sick. If you have a garden or crops, be sure to dig your pit as far away from those as possible to avoid contamination.

Dig a narrow pit that is 3 to 5 feet deep. Make it wide enough to use (about 4 to 6 inches in diameter) but not so wide that people or animals can fall in. After doing your business in the hole, wipe and throw the toilet paper in there, too (preferably the biodegradable kind), and cover the waste with a good amount of dirt. Place a board that completely covers the hole over the pit. The board should be thick and study enough that people can walk over it without it caving in. Be sure to keep the pit covered at all times when not in use. Properly covering your pit is critical—you don't want your pets getting into it. Remember that flies and other pests are attracted to fecal matter and allowing them access to it can result in a health nightmare for you and your family. Don't give them an opportunity to spread pathogens around.

When the pit is full, fill it in completely with dirt. If necessary, follow the same procedure to dig another pit. Dig all of your pits far apart, if possible.

As always, planning and preparing for these kinds of situations in advance is smart. It will save you time, trouble, and inconvenience. Using the methods above to manage and dispose of human waste is not ideal and must be used with caution to avoid contracting illnesses from contamination.

To avoid having to resort to using buckets, bags, and pits as toilets, consider purchasing camp toilets to have on hand for emergencies.

If you live in an area that is prone to hurricanes, flooding, or other natural disasters and you have the space and privacy, you may want to consider building an outhouse on your property. Outhouses are also fairly easy and inexpensive to make. And, because they don't use water, they can save you loads of money on your water bill.

Who knows, maybe you'll find that using an outhouse isn't such a bad thing at all and will end up using it on a regular basis. No? Well, fair enough. But it sure would be good to have one just in case of an emergency, right?

Before rushing out to build an outhouse on your property, be sure you are allowed to have one in the first place. Some regions have laws against such things, and some have certain restrictions and guidelines you'll have to comply with.

Once you get the go-ahead, here's how to build your own outhouse, complete with an iconic crescent-moon-shaped window.

First, choose a location. You'll want it as far away from water sources and animal pens as possible. Avoid any underground pipes or wiring, too. Check with your local utility service providers, if applicable, to find out where those lines are located.

Once you find an appropriate building spot, you'll need to gather these supplies:

- Shovel or backhoe
- Tape measure
- Post hole digger
- Four 8-foot long 4x4 wooden posts
- 1 50-lb bag concrete mix
- Water
- Miter saw, table saw, or circular saw (for cutting plywood)
- Five ½-inch-thick plywood sheets
- Small box of 2-inch wood screws
- Power screwdriver
- Metal panels, roll roofing, or shingles (enough to cover 5 foot square roof)
- Wooden planks (optional)
- Paint or stain (optional)
- Hinge fixtures
- Outhouse door
- Toilet seat
- Pencil
- Jig saw

Steps:

1. Dig a hole for the outhouse waste. Use a shovel or backhoe and a tape measure to dig a 3- to 5-foot-deep hole that is 2 feet square in width and length.
2. Dig four 2-foot-deep post holes for the supporting beams of the outhouse structure using a post hole digger. Mark off a 4-foot square on the ground around the 2-foot square hole, keeping that larger hole centered on the backside of the outhouse. Dig the post holes in the corners of the marked-off square.
3. Set the wooden posts into the post holes. Mix (using the amount of water indicated on package instructions) and pour concrete into the openings around the post holes to secure the posts.
4. Cut plywood pieces to cover the outhouse walls. Cut the first plywood piece 4 feet wide by 6 feet long and attach this piece to the back side of the outhouse. Screw the plywood piece into place on the 4x4 posts.
5. Cut two matching plywood pieces that measure 4 feet wide by 6½-feet long for the sides of the outhouse. On one side of each plywood piece, measure down 6 inches and make a mark with your pencil. Use this mark as a reference point to cut a slope into these pieces that drops 6 inches from one side to the other. Attach these pieces to the sides of the

outhouse structure with the taller side of each piece facing the front of the outhouse.

6. Cut another plywood piece measuring 6½ feet long by 4 feet for the front of the outhouse. Before you attach the piece, cut out an opening on this piece for the outhouse door.
7. Cut a final plywood piece that measures 5 feet square to serve as the roof. You'll need to stand on a ladder to attach this piece. Drive wood screws down through the top plywood piece and into the edges of the plywood pieces that make up the sides, front, and back of the outhouse.
8. Paint the exterior of the plywood outhouse structure or attach wooden planks to the outside walls of the outhouse for a more attractive exterior if you'd like. Apply roof panels or shingles to the top of the plywood roof piece to extend the life of the roof and prevent leaking.
9. Purchase or build a door that measures 2½ feet wide by 6 feet tall and set it into place at the front of the outhouse structure. Use hinges to attach the door to the outhouse; be sure to use hinges approved for the weight of the door you install. Cut a crescent moon shape into the top of the door if you'd like. This will allow some light to enter the outhouse and will also help some odors escape.

10. Construct the interior seat of the outhouse by building an open-ended 4-foot-wide by 2½-foot-wide box from plywood. Set the open end of the box over the hole. Place a premade toilet seat over the box where you wish to place the seat permanently and trace the opening. Use a jig saw to cut out this opening and then attach the toilet seat permanently with hinges to the top of the box shape.

Note: To improve ventilation (and reduce odors), put vent pipes in the corners of the outhouse (make sure they extend above the top of the outhouse roof, down through the bench, and into the pit), and insert screened windows at the top and bottom of the outhouse.

OUTHOUSE TIPS, MAINTENANCE, AND SAFETY

For lighting, install battery-operated push lights or motion-sensor lights inside of your outhouse.

Store toilet paper (preferably biodegradable) in sealed plastic bags or containers inside the outhouse to keep it dry, clean, and pest-free.

Place a bucket of sawdust, wood ash, or cedar shavings with a scooper in the outhouse. Sprinkle a bit of this into the hole each time the outhouse is used.

The use of lime and other odor-eating products to control smells is an option, but many experts say that such chemicals simply mask bad smells—they don't eliminate the cause. That's because they can actually slow down the decomposition process by interfering with the natural balance of oxygen and good bacteria. If you decide to use lime, be careful. It can burn skin, so don't get it on your hands or the outhouse seat.

If you build a vent system in your outhouse, use a toilet seat with a lid. Keeping the lid closed at all times will force odors to escape via your vent system and should help minimize bad smells.

Do NOT pour bleach or any products that contain chlorine into the hole. Chlorine can react with urine to create dangerous (and potentially fatal) ammonium chloride gas.

Using pesticides in an outhouse is not recommended. Instead, hang fly paper strips or traps inside.

Don't throw anything that is nonbiodegradable in the hole. Place a covered trash receptacle in the outhouse for those items.

Do NOT drop matches or cigarette butts (even if they are extinguished) into the pit because methane gas from waste breakdown may be in the hole. Methane can cause a fire or explosion.

NEVER allow children to use an outhouse by themselves.

IF you drop anything into the hole, don't try to retrieve it. It's not worth the risk. Contact a professional if the item is something you absolutely need to recover.

There will likely come a time when you discover that your outhouse hole is full. You will need to either clean it out yourself, contact a sanitation company to clean it, or move your outhouse to a new location.

ADDITIONAL SOURCES

CDC.gov/healthywater/emergency/cleaning/sanitation.html

DisasterStuff.com/store/pc/How-to-make-an-Emergency-Toilet-98p1134.htm#.VLgG1yvF-So

SewerSmart.org/images/SewerSmart_brochure.pdf

TacticalIntelligence.net/blog/survival-sanitation-how-to-deal-with-human-waste.htm

Toiletology.com/Disaster-toilet.shtml

HANDLING GARBAGE DURING AN EMERGENCY

If a natural disaster or some other event in your area causes trash services to cease, you'll need to use alternate means to manage garbage that accumulates. Handle trash the wrong way and you may end up getting sick or contaminating your water supply.

As soon as the grid goes down, start thinking about how you are going to manage the different kinds of trash your family generates. Then, develop a simple system to deal with it.

Begin with separating your trash based on what kind of items you are throwing out. Can any of the items be recycled or reused? Are any of the things you'll be discarding toxic or dangerous to the environment? Do any of them attract pests? Can any of your garbage be used for composting?

It is far easier (and tidier) to separate your garbage as you are discarding it instead of going through big bags later and trying to sort everything. This will also save you time and will help keep you healthy. A down-grid situation brings enough problems, and the last thing you need is for you or a family member to get sick from handling garbage.

Generally, there are five categories of garbage that most households generate:

1. Things that biodegrade quickly (plant and animal matter)
2. Paper products
3. Plastics and metals (these should be flattened or crushed to reduce bulk)
4. Sanitary items (diapers, feminine hygiene products, and the like)
5. Raw or cooked meat or other animal products

Using five trash cans or heavy-duty garbage bags, designate one for each of the categories. Label each with what should be placed in that receptacle.

According to TacticalIntelligence.net, here's how to handle what accumulates.

BIODEGRADABLE WASTE

Plant, fruit, and vegetable waste, including scraps and peels, can be disposed of by making compost piles. You should not mix plant and vegetable waste with animal manure. To build a compost pile for plant waste:

- Put plant and vegetable waste in a pile.
- Turn the pile over two or three times a week with a large shovel. If you own a composting bin, that's even better. Some of them spin, which provides you with an easy method to stir the compost. Place your plant and vegetable waste in the bin.
- When the plant material is reduced to a fine, dark-colored soil, it is ready to be used in your garden.

Some other items that can be composted are:

- Coffee grounds and filters
- Tea bags
- Eggshells
- Cooked rice, pasta, and oatmeal
- Paper towels, napkins, and plates
- Stale crackers and chips

If you are not able to go outside due to dangerous conditions, you can store these items in a container with a lid or a heavy-duty garbage bag. When you are able to go outside again, you can make your composting piles.

Note: Animal waste should not be kept inside your residence and should be stored separately from the waste listed above. Place it in sealed containers or bags until you can move it if you absolutely cannot go outside. Do not store animal waste near your water supply.

Because it is high in phosphorous and nitrogen, composted animal waste is a good ingredient for fertilizer. However, the waste from carnivorous animals should not be used to fertilize food crops. Adding lime to the waste can neutralize the flammability of the nitrogen in the waste. Lime also inhibits pathogens by raising the pH levels to a point where pathogens cannot survive.

PAPER PRODUCTS

Newspapers, cardboard, food packaging, and the like can usually be recycled, but if it is accumulating too rapidly for you to store it, you may have to burn it. Only burn items if you are sure conditions allow and the items are flammable. You'll need to use a large barrel or pit that is designed for burning items.

Your main priority here is keeping this kind of garbage away from your water supply, especially items that have food residue on them as they tend to attract pests and animals.

PLASTICS AND METAL

Store these items in a sealed bag or container away from your water supply until they can be properly recycled or discarded. Again, the concern here is keeping pathogen-carrying pests and animals away.

SANITARY ITEMS

Used diapers and feminine hygiene products can carry pathogens and will need to be handled carefully. Discard them in sealed containers or bags until your garbage service is restored. Another option is burying the items in a pit that is at least 3 feet deep and at least 100 feet from all water sources.

RAW OR COOKED MEAT OR OTHER ANIMAL PRODUCTS

Raw or cooked meat waste will draw pests and animals and become a health risk as it decomposes, so you'll need to deal with it in a safe and timely manner. DO NOT put these items in your garbage disposal. Meat and other animal products can contaminate a region's water supply if disposed of this way.

Your best bet is to dig a hole that is at least 4 feet deep (and at least 100 feet away from water sources) and bury the meat. Place the meat, poultry, or fish in the hole and sprinkle a bit of lime on it. Fill the hole in with dirt.

If you can't get outside or don't have a place to dig a hole, put the trash in a sealed bucket with a splash of bleach, a layer of wood ashes, or any other odor controlling substance that is inhospitable to bacteria and fungus until your garbage services are restored or until you can drive to a dump.

DISINFECTING YOUR WELL

If you have a well, the Centers for Disease Control and Prevention recommends disinfecting it after a flood or other natural disaster.

Wait until it is safe to go outside to do this. If your well or pump is run by electricity, disconnect it or turn it off prior to approaching the well to avoid being shocked. If there are downed power lines on your property or anywhere near your well, do not touch them. Do not attempt to use or clean your well until officials have repaired the power lines and you are sure the area is safe.

Protect yourself by wearing boots, goggles, and gloves while working outside after a disaster. Check carefully for hazards around the area, especially any glass, metal, or wood debris. Even if your well is intact and does not appear to be damaged, assume that it has been contaminated after a natural disaster, especially if there was flooding. Floodwaters can carry an alarming variety of pathogens and microbes, including bacteria, viruses, and protozoa. Exposure to these bugs can cause illnesses ranging from mild gastritis to serious diseases such as dysentery, infectious hepatitis, and severe gastroenteritis.

SHOCK CHLORINATION

Shock chlorination is the most widely recommended way to treat possible contamination of wells after a natural disaster.

If you have a drilled well with a hand pump, you can proceed with this method, but if your well has any electrical components, you will need to avoid cleaning your well until you are certain it is safe. An electrician, well contractor, or pump contractor can inspect your well and make this determination for you. They also can repair any electrical components that are damaged.

Excessive silt, sediment, and mud can enter wells during floods and major storms, and sometimes this requires professional help to remedy.

Licensed well drillers are trained and equipped to shock chlorinate wells, but if you'd like to do it yourself, here's how.

HOW TO SHOCK CHLORINATE A DRILLED WELL

There are certain safety precautions you should take before attempting to use this method to disinfect your well. Concentrated chlorine solutions for shock chlorination can be dangerous to handle. Wear protective gear including goggles, rubber gloves, and boots, and mix the chlorine solution in a well-ventilated area. Because the level of chlorine in your well water will be high after shock chlorination, you'll need to use an alternative source for drinking water.

Chlorine should have enough contact time to kill any bacteria lurking in your well. Make sure that no one in your home uses the water for any purpose during the 12- to 24-hour treatment.

Wait at least 7 to 10 days after shock chlorinating your well to test the water for total coliform and E. coli bacteria. Follow sample collection instructions carefully. If the test results show the absence of coliform bacteria, the water is safe to drink.

Use plain, unscented bleach with at least 5 percent sodium hypochlorite.

Follow these steps:

1. Start the pump and run water until it is clear. Use the outside faucet nearest to the well to drain the potentially contaminated water from the well and keep the unsafe well water out of the interior household plumbing. If no pump is installed, bail water from the well with a bucket or other device until the water is clear.
2. Use the formulas below to determine the amount of liquid household bleach needed. Three pints should be added for every 100 gallons of water in the well.
3. Determine the depth of water in your well, which is the distance from the bottom of the well to the water level. If you don't already know what the depth of your well, you can calculate it by measuring the distance from the ground level to the water level ("b").

4. Subtract "b" from the well depth "a" to find the total depth of the water: a - b = c. If you do not know the depth of your well, but you know the well drilling company who constructed it, contact them. Well drillers usually keep records of all the wells they drill. If you can't find any records about your well, you can contact a licensed well driller to assist you in taking the appropriate measurements.
5. Determine your well's storage per foot of water. This number is based on the diameter of your well. The inside diameter of the pipe of a drilled well is typically between 4 and 10 inches.

   Refer to the following table to determine your well's storage per foot of water.

Diameter of pipe (in inches)	Storage per foot of water (gallon/foot)
4	0.653
5	1.02
6	1.47
7	2.00
8	2.61
9	3.30
10	4.08

6. Multiply your total depth of water, "c," by your storage per foot of water, "s."

7. Pour 3 pints of bleach into your well for every 100 gallons of water it holds. (For example, if the volume of water in your well is 300 gallons, you will add 9 pints of bleach to treat the well.)

   Using a 5-gallon bucket, mix the bleach with 3 to 5 gallons of water.

9. Remove the vent cap. Pour the bleach water mixture into the well using a funnel. Avoid all electrical connections. Attach a clean hose to the nearest hose bib and use it to circulate water back into the well to thoroughly mix it.
10. Rinse the inside of the well casing with a garden hose or bucket for 5 to 10 minutes.
11. After at least 12 hours, attach a hose to an outside faucet again and drain the chlorinated water onto a non-vegetated area, such as a driveway. Continue draining until the chlorine odor disappears. Avoid draining into open sources of water (streams, ponds, etc.).
12. Until your well water has been tested, filter it and then boil it (rolling boil for one minute) before use, or use an alternative water source. Wait at least 7 to 10 days after disinfection, then have the water in your well tested. Water testing cannot be done until all traces of chlorine have been flushed from the system.

13. Test the water for total coliform and either E. coli or fecal coliform bacteria to confirm that the water is safe to drink.

    If results show no presence of total coliform or fecal coliform, the water will be considered safe to drink, at least from a microbial contamination standpoint.
14. Follow up with two additional samples, one in the next two to four weeks and another in three to four months.

CHAPTER THIRTEEN

WATER CONSERVATION

We've talked about doing without water. We've talked about unsafe water. We've talked about storing it, testing it, purifying it, and collecting it. But one thing that people don't always stop to consider is exactly how much water they use in a day.

If you have to go to the effort of obtaining it, purifying it, getting it into your house, and heating it, you don't want to waste a drop. There's no better way to become an expert at conservation than to haul buckets of water for your personal use. This I know from arm-aching experience.

Everyone in the preparedness realm knows the adage about 1 gallon per person per day, but that is only the tip of the iceberg. It doesn't include the vast amount of water we customarily use for hygiene purposes. The average American goes through at least 100 gallons of water per day. We take long, hot showers. We run the tap when we brush our teeth. We think nothing of running the dishwasher or throwing a load of laundry into the washing machine each day.

Clearly, in an off-grid scenario, many of those activities wouldn't be possible. In a situation where water doesn't flow from the taps, you'll be responsible for hauling it into the house by the bucketful. If you want hot water, you'll have to use an off-grid method to heat it. If you're lucky, you'll have a water tank warmed by the sun. If you aren't so lucky, you'll be heating it over a fire.

There's also the possibility of a slightly different situation. Perhaps your water supply is rationed and limited by the public utility companies due to the terrible drought sweeping most of the United States. You're still going to want clean clothes, clean dishes, and a clean body. You'll want to be able to flush your toilet without using half of your day's "ration" of water.

No matter why your water supply is limited, you'll want to conserve that precious fluid.

IF YOU STILL HAVE RUNNING WATER

You may still wish to conserve when the luxury of running water is available. Perhaps, as mentioned above, water has been rationed by your local utility company. Your well might be low, or maybe you just want to be environmentally responsible and reduce your usage.

- **Reduce the amount of water per flush**. Use a brick, a filled plastic bottle, or a float booster to fill space in the back of the toilet tank. This allows the tank to fill with a smaller amount of water during each flush.
- **Don't flush every time**. Speaking of flushing, you may have heard the rhyme, "If it's yellow, let it mellow. If it's brown, flush it down."
- **Reuse water that would normally go to waste.** Devise a graywater catchment system for your shower, washing machine, and kitchen. This water can be used for flushing, watering plants, and cleaning.
- **Take shorter showers**. Try to reduce your showers to five minutes. This can save up to 1,000 gallons per month! If you can't handle a five-minute shower, just try to make them a little quicker. For every two minutes you shorten your shower time, you can save approximately 150 to 200 gallons per month. So if you can't quite handle a five-minute shower, you can still reduce your usage dramatically. For example, if your usual shower is 20 minutes and you reduce it to 10 minutes, you could still save up to 1,000 gallons of water per month from your previous usage.

- **Upgrade your shower head**. Install a water-saving shower head. A low-flow shower head uses 2 gallons of water per minute, whereas a conventional shower head uses 3.5 to 4 gallons of water per minute. This can result in a savings of more than 5,000 gallons of water per year, per person. So, for that family of four, this works out to a whopping 20,000 gallons of water from a simple change of hardware.
- **When you have a shower, plug the tub**. Use the water you collect for hand-washing laundry. (See the next suggestion!)
- **Handwash some of your laundry**. For delicate items or things that are lightly soiled, handwashing can save a lot of water, particularly if you use water that would have gone down the drain otherwise.
- **If you do use a dishwasher, run it only when it's completely full.** This can save you 1,000 gallons per month, since you'll most likely be skipping a day between loads.
- **Wash your dishes by hand**. Fill your sink with soapy water. You'll be using a fraction of the water it takes to run your dishwasher. (Bonus: You get to save on electricity, too!)
- **Run a basin of rinse water**. Instead of running water over each dish to rinse, fill one side of the sink or a basin with rinse water containing a splash of

white vinegar. When washing your hands, dip them in a basin of water, lather up, then rinse under running water. Running water uses up to 4 gallons per minute.

- **Don't be wasteful!** If you drop a tray of ice cubes, pop them into a pet dish or into your potted plants.
- **Upgrade all of your faucets**. Inexpensive aerators with flow restrictors greatly reduce the flow of water from your taps.
- **Use a nozzle on your hose**. Spray water only where you want it by placing a nozzle on your hose that allows you to direct the flow onto your plants or into the dog's dish, instead of spraying it uselessly as you walk across the lawn to the garden.
- **Repair leaky faucets**. The rate of one drip per second adds up to 5 gallons per day literally down the drain. Over the course of a year, that is nearly 2,000 gallons, a tiny drip at a time.
- **Check your toilet for leaks**. It isn't just your faucet that might be dripping away thousands of gallons of water per year. If your toilet is leaking you might never even notice it. Here's how to check: Put some food coloring in the tank. Don't flush. Just walk away for 30 minutes. When you come back, there should be no trace of color in the toilet bowl. If there is some color, you have a leak. Good news: most replacement parts to fix this are very inexpensive.

- **Stop using the garbage disposal**. You use a LOT of water running the disposal. Make better use of those food scraps by taking them out to the compost pile.
- **Shop wisely**. If you are buying new appliances and fixtures for your home, opt for those that use water more efficiently, like front-loading washing machines and low-flow toilets.
- **Reuse nutrient-rich "dirty water."** When you clean out your fish tank, reserve the water for your garden. Your veggies will love the nutrient boost!
- **Don't waste the water you run waiting for hot water**. One of the biggest water wasters is something we ALL do: we run the taps for a minute, waiting for the flow to get warm. Collect that water! You can use it for cleaning, for the pets, for watering plants, for making tea, or for cooking. Just don't let it go down the drain.
- **Same with cold water**. Another water waste occurs when you run the taps to get cold water for drinking. For cold water on demand, store a pitcher of water in the refrigerator instead.
- **Don't run the water while you brush your teeth**. You're sending gallons down the drain. Only run the water when you need to rinse your toothbrush or the sink.

IF YOU DON'T HAVE RUNNING WATER

Life without running water…ugh. Been there, done that, got a stain on my T-shirt. Luckily for me, it happened only briefly, due to power outages at the cabin. We were never completely without running water for more than a week. However, it prepared me because it helped me to understand just how much water we use and how much we rely on the faucets in our homes.

Do you remember the "No Running Water Drill" we did at the beginning of the book? You and your family did a test run to see how much water you used over the weekend. Now, imagine that weekend was extended into a week, a month, or an indefinite amount of time. Take it a step further—you're no longer using stored water. You're collecting it, hauling it into your house, purifying it, and heating it. That's a lot of work, so you won't want to waste a drop.

Many of the suggestions above can be modified to extend the amount of water available in a situation during which you have no running water, but for real inspiration on living without running water, the best advice is to look to the past.

My Granny grew up in a house without running water. They got their water by manually pumping it from a well and lugging it to the house from the water source. Even in her eighties, she was very thrifty with her use of water. Hauling buckets during her childhood made a life-long impression. They took many of the following steps in order to lessen the workload. Here are some things that I learned at Granny's house:

- **Reuse cooking water.** First of all, the water you used for cooking was already purified and made safe to consume. Don't waste it! If you have boiled pasta or vegetables, use this water for making soup. You will have retained some of the nutrients and flavor from the first thing you cooked in the water.
- **Use a cup for shaving.** This one is for the gentlemen. When shaving, you can make lather in one cup, then rinse your razor in another cup between swipes.
- **Use a glass of potable water for brushing your teeth**. Make sure you use purified water for brushing your teeth. It's unlikely you'll get sick, since you're spitting the water out, but there's no point in risking illness, especially during an emergency situation during which resources are limited. Finish up with some mouthwash.
- **Use a pitcher and bowl for washing up.** When we lived in our cabin, I got a pretty, old-fashioned pitcher and bowl to put on a stand outside of the bathroom. Look to the old-fashioned ways for inspiration in solving problems.

- **Wash clothes in your bathtub**. You can use soapy water from throughout the day for the first soak. (See page 159.)
- **Wash produce in a basin of water.** Don't forget to wash your produce! Chemicals, feces, and all manner of bacteria could be lurking. Wash in a basin with food-safe soapy water, then rinse in another basin. Save the water for other uses, like flushing.
- **Harvest rainwater for your garden**. Make sure you have a way to collect water to keep your garden growing. You'll also need a way to dispense it. Many water barrels have a gravity-fed spout to which you can attach a hose, sparing you from lugging buckets of water to your garden.
- **Use an organic mulch in your garden**. This will help to retain moisture, allowing you to water less frequently. This is a definite bonus if you are watering via buckets.
- **Water early or late in the day.** Don't water your plants during the hottest part of the day. Much of the water will evaporate. Instead, water early in the morning or at dusk.

A FEW MORE MODERN TIPS

Landscape with plants that grow naturally in your area. They should require little in the way of additional watering. On average, an astounding 60 percent of a household's water usage goes toward lawn and garden maintenance.

Grow organic. Chemical fertilizers can increase a plant's need for water.

To flush or not to flush? If you have a septic system, you can still flush waste (see page 164). This is such a step up in a down-grid situation. However, you still have to do some things manually, since water won't be pumped to your toilet. Simply add water to the tank when it's "brown" (remember the rhyme about flushing?). This is the perfect use for your graywater from dishes, showers, tooth-brushing, and laundry. It's getting flushed so it doesn't matter if the water is gunky.

CONCLUSION

Did you ever expect that there could be so much information about water? When I began researching this project, I wondered how on earth I could come up with enough content for an entire book. As I immersed myself in the topic, I began to wonder instead how I could condense the reams of information into a manageable package.

Because it's an overwhelming amount of information, and such a large amount of it is technical, this last section will summarize each chapter. You can refer back to the full chapter for detailed information on each topic as needed.

Chapter One: Even Non-Preppers Need Water Whether you're on board with the idea of being a prepper or not, it's difficult to deny that water storage is vitally important. Many unpleasant situations have the potential to become life-threatening in a matter of days due to a lack of potable water.

This chapter covers such situations that have occurred in the US and across the ocean, in places we consider less modern. As this chapter highlights, a sudden water crisis can strike anywhere, at any time. It's up to you to be prepared or face the consequences.

Chapter Two: Dehydration Can Be Deadly Dehydration is common during a disaster situation, and it can be deadly. This chapter highlights major causes, symptoms, and treatment options.

Chapter Three: Toxins in Municipal Water Supplies Most people believe that the water flowing from their taps is safe, particularly if it comes from the local water treatment facility. Unfortunately, this isn't always true. Pollution from many different sources has affected our water supply.

Depending on where you live, the water flowing through your taps and faucets likely comes from one of two sources—groundwater or surface water. There are problems inherent with each, as detailed in this chapter.

Chapter Four: Water-Related Illnesses The number one cause of death in the aftermath of a disaster is waterborne illness. When people are desperately thirsty, their need to quench their thirst overrides their awareness that the water they are drinking may not be safe. Chapter Four summarizes how water sources can become contaminated. Symptoms and treatment options for serious illnesses that can result from waterborne illnesses are also discussed.

Chapter Five: A Glimpse at Everyday Life Without Running Water In Chapter One, we discussed that an interruption in water service isn't just something that happens across the ocean.

When my daughter and I spent a year in a cabin in the woods of North Central Ontario, we dealt with regular interruption of running water. Our well had an electric pump, and when winter storms raged, power went out frequently. Chapter Five outlines the many simple lessons we learned to handle life without running water. We learned firsthand of the importance of figuring these things out before you are actually stranded for a week without water and power.

Chapter Six: Creating a Water Plan By now, you would have learned the risks inherent in an interruption of the water supply. It's time to make a plan. A four-layer plan, described in Chapter Six, will be the most likely way to cover all your bases.

Chapter Seven: Storing Water Now that you know how much water you need, it's time to start safely storing enough to get you through a two-week interruption. This chapter will get you there, guiding you through all sorts of containers and systems out there for storing water.

Chapter Eight: Acquiring Water Even if you've developed and followed a water plan, there's a chance your supply will run dry.

If this happens, you'll need to have a backup plan, which includes finding accessible and usable water sources. Don't wait until an emergency happens to figure this out. Chapter Eight will help you locate sources of water that you'll be able to collect, purify, and store.

Chapter Nine: Testing Your Water By now, you will be well aware of the countless pathogens and pollutants that can infect just about any source of water. Water is a universal solvent, after all, and this quality means that it is very easily contaminated.

At a bare minimum, you should test your water source for coliform bacteria and nitrates, because both can cause serious health problems and their presence often means other contaminants are in your water, too. Lead, arsenic, and fluoride are hidden dangers you'll need to check for as well. Chapter Nine will go over these important steps.

If you notice changes in the smell, taste, or color of your water, or if you have spilled any potential toxins near or in your well, test the water before consuming it. This chapter will guide you through the process, including options for inexpensive to comprehensive water-testing.

Spending a bit of time and money to check your water for dangers can provide you with peace of mind and protect your family's health—a worthy investment, indeed.

Chapter Ten: Purifying Water Water from most sources needs to be purified before consumption. There are many water filtration and purification methods to choose from.

Which method (or methods) you choose will depend on several factors, as described in Chapter Ten.

Chapter Eleven: Finding Emergency Sources of Water There may come a time when, despite having the best-laid plans, the unthinkable happens. You run out of water, and your backup sources are unavailable. Thankfully, with a little creativity and help from Chapter Eleven, you can find hidden sources of water in your home and on your property.

Chapter Twelve: Sanitation More people die after natural disasters than during them, and it is usually because of poor sanitation practices. It only takes one person or misstep to contaminate a water supply.

For that reason, it is important to know how to handle human waste and effectively wash and clean to protect water supplies from infection with dangerous pathogens. This chapter will show you how.

Chapter Thirteen: Water Conservation By now you know about the toxins in water. You also know how to store, test, purify, find, and collect it. Now you must learn how to conserve it.

Don't take this precious commodity for granted. In this chapter, you'll discover things you can do on a regular basis to conserve water, whether it's still streaming generously through your taps or has stopped running.

MASTER SUPPLY LIST

To make things a little easier, here's a list of supplies that I've recommended throughout the book. You don't need every single one of these items. Some of them are just different options to meet the same needs. Use this master list to create a personal checklist of supply requirements.

- ❑ 5-gallon buckets with lids (keeping several of these on hand is a good idea, 4 to 6 as a minimum)
- ❑ 5-gallon water jugs and dispensers
- ❑ Baby wipes and refills
- ❑ Baking soda
- ❑ Biodegradable toilet paper
- ❑ Borax
- ❑ Chlorine tablets, like Portable Aqua
- ❑ Cistern or large tank
- ❑ Clothes-drying racks
- ❑ Garbage cans with lids (several)
- ❑ Gatorade and/or Pedialyte
- ❑ Granular calcium hypochlorite
- ❑ Gravity filtration system, like those made by Berkey, Platypus, or Sawyer
- ❑ Hand sanitizer (with at least 60 percent alcohol)
- ❑ Heavy-duty garbage bags (lots!)

- ❑ Hydrogen peroxide
- ❑ Indoor trash cans with lids (several)
- ❑ Iodine tablets or Polar Pure (iodine crystals)
- ❑ Janitor's mop with a press wringer (good quality)
- ❑ Kitty litter
- ❑ Laundry brush
- ❑ Laundry soap
- ❑ Non-scented liquid household bleach with chlorine
- ❑ Pails and buckets (various sizes)
- ❑ Plunger
- ❑ Portable water cooler with spigot
- ❑ RainSaucers
- ❑ Rainwater-harvesting supplies: storage barrels or urns, downspouts, and leaf screens
- ❑ Reverse osmosis or distillation system, or both
- ❑ Tarp or plastic sheeting (to make a solar still)
- ❑ Ultraviolet (UV) light water purifier (should be used in conjunction with reverse osmosis)
- ❑ WaterBOB
- ❑ WaterBrick storage system
- ❑ Water-testing kits
- ❑ Washboard
- ❑ Wonderwash

Now, get busy!

Wouldn't it be a shame to have acquired all of this information, but not have put any of it into practice before disaster strikes? We don't get a two-day warning. Now is the time to prepare for your family's survival.

RESOURCES

WEBSITES:

TheOrganicPrepper.ca

UndergroundMedic.com

ReadyNutrition.com

SHTFplan.com

GrayWolfSurvival.com

PrepperWebsite.com

GOVERNMENT AGENCIES:

Fema.gov

CDC.gov

Who.int/en

BOOKS:

The Pantry Primer: A Prepper's Guide to Whole Food on a Half Price Budget

The Prepper's Blueprint

Rainwater Harvesting and Use: Understanding the Basics of Rainwater Harvesting

PRODUCTS:

BerkeyFilters.com

RainSaucers.com

WaterBob.com

Sawyer.com/products/sawyer-mini-filter

INDEX

ACKNOWLEDGMENTS

Without the help of my friend Lisa Egan, this book would not be possible. She served as a science consultant and research analyst throughout the process. The technical portions of this book are due to many hours of her input. Thanks, Lisa. GM!

Thank you to my friend Lizzie Bennett of *Underground Medic* for your input on the medical aspects of the book and for being around for the long haul. Thank you to my friend Tess Pennington for allowing me to use an excerpt from your book, *The Prepper's Blueprint*, and for your unflagging support over the years. And finally, thank you to my friend and mentor Mac Slavo, for the gentle push out of the nest. It was just what I needed. Okay, you actually shoved me, kicking and screaming, but I still appreciate it.

ABOUT THE AUTHOR

Daisy Luther is an author and blogger who lives in a small village in the foothills of Northern California. She is the author of *The Organic Canner* and *The Pantry Primer: A Prepper's Guide to Whole Food on a Half Price Budget*. Her website is *The Organic Prepper*, and she is the cofounder of the website *Nutritional Anarchy*. Daisy uses her background in alternative journalism to provide a unique perspective on health, nutrition, and preparedness. Daisy's articles are widely republished throughout alternative media. She is the proud mom of two wonderful and talented daughters.